CAMBRIDGE LIBRARY COLLECTION

Books of enduring scholarly value

Mathematical Sciences

From its pre-historic roots in simple counting to the algorithms powering modern desktop computers, from the genius of Archimedes to the genius of Einstein, advances in mathematical understanding and numerical techniques have been directly responsible for creating the modern world as we know it. This series will provide a library of the most influential publications and writers on mathematics in its broadest sense. As such, it will show not only the deep roots from which modern science and technology have grown, but also the astonishing breadth of application of mathematical techniques in the humanities and social sciences, and in everyday life.

Dynamics

A.S. Ramsey (1867-1954) was a distinguished Cambridge mathematician and President of Magdalene College. He wrote several textbooks 'for the use of higher divisions in schools and for first-year students at university'. This book on dynamics, published in 1929, was based upon his lectures to students of the mathematical tripos, and reflects the way in which this branch of mathematics had expanded in the first three decades of the twentieth century. It assumes some knowledge of elementary dynamics, and contains an extensive collection of examples for solution, taken from scholarship and examination papers of the period. The subjects covered include vectors, rectilinear motion, harmonic motion, motion under constraint, impulsive motion, moments of inertia and motion of a rigid body. Ramsey published a companion volume, Statics, in 1934.

Dynamics

A Text-Book for the use of the Higher Divisions in Schools and for First Year Students at the Universities

ARTHUR STANLEY RAMSEY

CAMBRIDGE
UNIVERSITY PRESS

CAMBRIDGE UNIVERSITY PRESS

Cambridge New York Melbourne Madrid Cape Town Singapore São Paolo Delhi

Published in the United States of America by Cambridge University Press, New York

www.cambridge.org
Information on this title: www.cambridge.org/9781108003148

This edition first published 1929
This digitally printed version 2009

ISBN 978-1-108-00314-8

DYNAMICS

Cambridge University Press
Fetter Lane, London

New York
Bombay, Calcutta, Madras
Toronto
Macmillan

Tokyo
Maruzen-Kabushiki-Kaisha

DYNAMICS

A Text-Book for the use of the
Higher Divisions in Schools
and for
First Year Students at the Universities

by

A. S. RAMSEY, M.A.

*President of Magdalene College,
Cambridge; and University Lecturer
in Mathematics*

CAMBRIDGE
AT THE UNIVERSITY PRESS
1929

PREFACE

This book is intended primarily for the use of students in the higher divisions in schools, particularly for those who intend to take an Honours Course of Mathematics at a University, and also for University students preparing for a first Honours Examination. It is based upon courses of lectures given during many years to first-year students preparing for the Mathematical Tripos, and it is assumed that the majority of readers will already have acquired some knowledge of elementary dynamics. Although the book contains chapters on Orbits and the dynamics of Rigid Bodies, none the less it may claim to be a text-book on *Elementary* Dynamics, for there is probably no branch of elementary Mathematics the content of which has expanded so greatly in the last twenty years.

One of the changes that accompanied the reform of the Mathematical Tripos was the removal of the restriction that Elementary Mechanics meant Mechanics without the Calculus. This restriction set well-defined and narrow bounds to the subject and the new regulations which gave teachers and students freedom to use any analytical methods in their work have been far reaching in their effect. Though the schedule in Dynamics for Part I of the new Tripos has remained unaltered, successive Examiners have added considerably to the interpretation of its contents. To give one instance only—the phrase 'motion under gravity' is now understood to mean 'in a resisting medium'—and it would be easy to give other examples of the elasticity of interpretation to which the schedule lends itself. The result of this change is that a first-year course in Dynamics at the University now includes all the easier problems of two-dimensional dynamics stopping short of the use of moving axes and Lagrange's Equations. This growth in the content of Elementary Dynamics has been a gradual process and undoubtedly beneficial to the study of the subject and stimulating to the average student. It is inevitable that its effect will extend to the schools, if it has not already done so; and it is not unreasonable to suppose that before many years have passed, candidates for Scholarships in Mathematics will be expected to possess a wider knowledge of dynamics embracing such parts of the subject as 'motion under simple central forces' and the elements of uniplanar rigid dynamics. The object of this book is to assist in this development. It is hoped that the presentation of the subject will prove sufficiently simple. An attempt has been made to preserve the conciseness of lecture notes and at the same time to give detailed explanations where experience has shewn that students find difficulties. Besides examples for solution the book contains a large number of worked examples; some of these are of purpose very simple illustrations of the theory, while others are of a more difficult kind for the assistance of readers who wish to learn how to work harder examples. The examples are nearly all taken from Scholarship papers or Tripos papers and the source is indicated by the letters S. and M. T. No attempt is made to exhaust the subject and the later chapters are only intended to be suggestive of the kinds of problems that can be solved, without elaborate analysis, as examples of the fundamental theorems; some few of these may prove to be too difficult for weaker students and they are intended rather to introduce abler students to more advanced work.

In conclusion I desire to express my thanks to the printers and readers of the University Press for their excellent work in the setting up of the book and the elimination of mistakes, and also to say that if the book contains errors I shall be grateful to anyone who will point them out.

A. S. RAMSEY

30 *Nov.* 1928
CAMBRIDGE

CONTENTS

Chapter I: INTRODUCTION

Chapter II: VECTORS

Chapter III: RECTILINEAR MOTION. KINEMATICS

Chapter IV: RECTILINEAR MOTION. KINETICS

DYNAMICS

Chapter I

INTRODUCTION

1·1. The subject of Dynamics is generally divided into two branches: the first, called Kinematics, is concerned with the geometry of motion apart from all considerations of force, mass or energy; the second, called Kinetics, is concerned with the effects of forces on the motion of bodies.

1·2. In order to describe the motion of a body or of a point two things are needed, (i) a frame of reference, (ii) a time-keeper. It is not possible to describe absolute motion, but only motion relative to surrounding objects; and a suitable frame of reference depends on the kind of motion that it is desired to describe. Thus if the motion is rectilinear the distance from a fixed point on the line is a sufficient description of the position of the moving point; and in more general cases systems of two or of three rectangular axes may be chosen as a frame of reference. For example, in the case of a body projected from the surface of the Earth a set of axes with the origin at the point of projection would be suitable for the description of motion relative to the Earth. But, for the description of the motion of the planets, it would be more convenient to take a frame of axes with an origin at the Sun's centre.

1·3. It is important to realize that there is no such thing as absolute time, but the period of rotation of the Earth relative to the fixed stars provides a unit of time, *the sidereal day*, which, so far as it can be tested with other time measures, is constant and therefore adequate for the purposes of ordinary dynamics.

1·4. The functions involved in dynamical problems are for the most part differential coefficients with regard to 'time,' '*t*,' as the independent variable. Thus 'motion' is 'change of position' or 'displacement,' 'velocity' is 'rate of displacement' and 'acceleration' is 'rate of change of velocity.' Hence, if x denotes a distance, dx/dt denotes a velocity and d^2x/dt^2 denotes an acceleration. The formulation of a dynamical problem therefore in general consists of one or more relations between certain variables (coordinates of position) and their differential coefficients with regard to time. Such relations are called differential equations.

NOTE ON DIFFERENTIAL EQUATIONS

1·5. It is assumed that the reader is acquainted with the elementary processes of differentiation and integration.

A *differential equation* is a relation between an independent variable t, a dependent variable x, and one or more of the differential coefficients of x with regard to t. The *order* of a differential equation is that of the highest differential coefficient that it contains. A *solution* of a differential equation is a relation between x and t that satisfies the equation, and the *complete solution* of a differential equation is a relation between x, t and one or more arbitrary constants of integration, the number of such constants being equal to the order of the equation.

For example:

$$\text{(i)} \quad \frac{dx}{dt} - 2x = 0$$

is a differential equation of the first order. It will be found on substitution that $x = e^{2t}$ is a solution; and the complete solution is $x = Ce^{2t}$, where C is an arbitrary constant.

$$\text{(ii)} \quad \frac{d^2x}{dt^2} + x = 0$$

is a differential equation of the second order. It has solutions

$$x = \sin t \text{ and } x = \cos t,$$

and the complete solution is

$$x = A \sin t + B \cos t,$$

where A and B are arbitrary constants.

1·6. The differential equations of dynamics are of either the first or second order.

Equations of the First Order.

We may have to deal with equations in which the variables can be separated. Such equations can be put in the form

$$M\,dx/dt = N \quad \dots\dots\dots\dots\dots\dots\dots(1),$$

where M is a function of x only (or a constant) and N is a function of t only (or a constant). The complete solution is

$$\int M\,dx = \int N\,dt + C \quad \dots\dots\dots\dots\dots\dots(2),$$

where C is an arbitrary constant.

For example, the equation

$$x\frac{dx}{dt} = g - kx^2$$

is solved by writing

$$\frac{x\,dx}{g - kx^2} = g\,dt,$$

so that

$$-\frac{1}{2k}\log(g - kx^2) = gt + C$$

is the complete solution.

1·61. Another type of equation that sometimes occurs in dynamics is the *linear equation of the first order*. A differential equation is said to be linear when it does not contain powers or products of the dependent variable x and its differential coefficients. Thus the linear equation of the first order is

$$dx/dt + Mx = N \quad \dots\dots\dots\dots\dots\dots\dots(3),$$

where M, N are functions of t or constants.

The solution is effected by first multiplying both sides of the equation by $e^{\int M\,dt}$ and then integrating; because it can easily be verified that

$$\frac{d}{dt}(xe^{\int M\,dt}) = \left(\frac{dx}{dt} + Mx\right)e^{\int M\,dt}$$

Hence

$$xe^{\int M\,dt} = \int e^{\int M\,dt}N\,dt + C\dots\dots\dots\dots\dots\dots(4),$$

where C is an arbitrary constant.

We note that if M is a constant the solution is

$$xe^{Mt} = \int e^{Mt}N\,dt + C \quad \dots\dots\dots\dots\dots\dots(5).$$

For example, the equation

$$\frac{dx}{dt} + kx = gt$$

can be integrated if both sides are multiplied by e^{kt}, giving on integration

$$xe^{kt} = g\int e^{kt}t\,dt + C$$

$$= ge^{kt}\left(\frac{t}{k} - \frac{1}{k^2}\right) + C,$$

or

$$x = \frac{g}{k}\left(t - \frac{1}{k}\right) + Ce^{-kt}\dots\dots\dots\dots\dots\dots(6).$$

1·7. Equations of the Second Order.

A common type of differential equation of the second order is

$$\frac{d^2x}{dt^2} + 2a\frac{dx}{dt} + bx = 0 \quad \dots\dots\dots\dots\dots\dots(7),$$

where a and b are constants.

It is easily seen by substitution that $x = e^{mt}$ is a solution of this equation, provided that

$$m^2 + 2am + b = 0 \quad \dots\dots\dots\dots\dots\dots(8),$$

so that, if m_1, m_2 are the roots of this quadratic, the complete solution of (7) is

$$x = C_1 e^{m_1 t} + C_2 e^{m_2 t} \quad \dots\dots\dots\dots\dots\dots(9),$$

where C_1, C_2 are two arbitrary constants.

The roots of (8) are $-a \pm \sqrt{(a^2 - b)}$, and there are three cases to be considered:

(i) *Real roots.* $a^2 - b = n^2$, say; then (9) may be written

$$x = e^{-at}(C_1 e^{nt} + C_2 e^{-nt}) \quad \dots\dots\dots\dots\dots(10).$$

(ii) *Equal roots.* $a^2 - b = 0$. The form $e^{-at}(C_1 + C_2)$ is inadequate for a *complete* solution, since $C_1 + C_2$ can only be regarded as one arbitrary constant, and, since the differential equation is of the second order, the complete solution should contain two. It is easily verified however that, when $a^2 = b$, the form

$$x = e^{-at}(C_1 + C_2 t) \quad \dots\dots\dots\dots\dots\dots(11)$$

satisfies equation (7), and since it contains two arbitrary constants, it is the complete solution.

(iii) *Imaginary roots.* $a^2 - b^2 = -n^2$, say; (9) may now be written

$$x = e^{-at}(C_1 e^{i nt} + C_2 e^{-i nt}),$$

or $\qquad\qquad x = e^{-at}\{(C_1 + C_2)\cos nt + i(C_1 - C_2)\sin nt\};$

which again may be written in the more convenient form

$$x = e^{-at}(C \cos nt + C' \sin nt) \quad \dots\dots\dots\dots(12).$$

Special cases of the foregoing. When $a = 0$.

(a) The complete solution of

$$\frac{d^2 x}{dt^2} - n^2 x = 0 \quad \dots\dots\dots\dots\dots\dots(13)$$

is $\qquad\qquad\qquad x = A e^{nt} + B e^{-nt}$

or $\qquad\qquad\qquad x = C \cosh nt + D \sinh nt \Big\} \quad \dots\dots\dots\dots(14),$

where A, B or C, D are the arbitrary constants.

(β) The complete solution of

$$\frac{d^2 x}{dt^2} + n^2 x = 0 \quad \dots\dots\dots\dots\dots\dots(15)$$

is $\qquad\qquad\qquad x = A \cos nt + B \sin nt$

or $\qquad\qquad\qquad x = C \cos (nt + a)$

or $\qquad\qquad\qquad x = C' \sin (nt + a') \quad \Bigg\} \quad \dots\dots\dots\dots(16),$

where A, B or C, a or C', a' are the arbitrary constants

1·71. Numerical Examples.

(i) $\dfrac{d^2x}{dt^2} - 4\dfrac{dx}{dt} + 3x = 0$; $x = Ae^t + Be^{3t}$.

(ii) $\dfrac{d^2x}{dt^2} - 4\dfrac{dx}{dt} + 4x = 0$; $x = e^{2t}(A + Bt)$.

(iii) $\dfrac{d^2x}{dt^2} - 4\dfrac{dx}{dt} + 13x = 0$; $x = e^{2t}(A\cos 3t + B\sin 3t)$.

(iv) $\dfrac{d^2x}{dt^2} - 4x = 0$; $x = Ae^{2t} + Be^{-2t}$.

(v) $\dfrac{d^2x}{dt^2} + 4x = 0$; $x = A\cos 2t + B\sin 2t$.

Chapter II

VECTORS

2·1. The physical quantities or measurable objects of reasoning in Applied Mathematics are of two classes. The one class, called **Vectors**, consists of all measurable objects of reasoning which possess directional properties, such as *displacement, velocity, acceleration, momentum, force*, etc. The other class, called **Scalars**, comprises measurable objects of reasoning which possess no directional properties, such as *mass, work, energy, temperature*, etc.

The simplest conception of a vector is associated with the displacement of a point. Thus the displacement of a point from A to B may be represented by the line AB, where the length, direction and sense (AB not BA) are all taken into account. Such a displacement is called a *vector* (Latin *veho*, I carry). A vector may be denoted by a single letter, e.g. as when we speak of 'the force P,' or 'the acceleration f,' or by naming the line, such as AB, which represents the vector. When it is desired to indicate that symbols denote vectors it is usual to *print* them in Clarendon type, e.g. **P**, and to *write* them with a bar above the symbol, e.g. $\overline{P}$, $\overline{AB}$.

Since the displacement from B to A is the opposite of a displacement from A to B, we write

$$\overline{BA} = -\overline{AB}$$

and take vectors in opposite senses to have opposite signs. Since two successive displacements of a point from A to B and from B to C produce the same result as a single displacement from A to C, we say that the vector AC is equal to the sum of the vectors AB, BC and write

$$\overline{AC} = \overline{AB} + \overline{BC} \quad \ldots\ldots(1),$$

and further, if $A, B, C \ldots K, L$ are any set of points

$$\overline{AL} = \overline{AB} + \overline{BC} + \ldots + \overline{KL} \quad \ldots\ldots\ldots\ldots\ldots(2).$$

Vectors in general are not localized; thus we may have a displacement of an assigned length in an assigned direction and sense but its locality not specified. In such a case all equal and parallel lines in the same sense will represent the same vector. On the other hand, vectors may be localized, either at a point, e.g. the *velocity* of a particle; or in a line, as for example a *force* whose line of action (but not point of application) is specified.

2·2. Composition of Vectors. A single vector which is equivalent to two or more vectors is called their **resultant**, and they are called the **components** of the resultant. Vectors are compounded by geometrical addition as indicated in formulae (1) and (2) of the last Article.

A vector can be resolved into two components in assigned directions in the same plane; for if AC be the vector, and through A, C two lines are drawn in the assigned directions meetings in B, then AB, BC are the components required.

When a vector is resolved into two components in directions at right angles to one another, each component is called the **resolved part of the vector** in the direction specified. Thus if a vector **P** makes an angle α with a given direction Ox, the resolved parts of **P** in the direction Ox and in the perpendicular direction Oy are

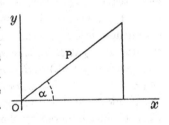

P cos α and **P** sin α.

2·3. Since the algebraical sum of the orthogonal projections on any straight line of the sides of a closed polygon is zero, it follows that the orthogonal projection of the resultant of a number of vectors is equal to the algebraical sum of the projections of the component vectors.

2·4. Analytical Method. To compound n vectors $\mathbf{P_1}, \mathbf{P_2}...\mathbf{P_n}$. Let the vectors make angles $\alpha_1, \alpha_2 ... \alpha_n$ with an axis Ox. Each vector may be resolved into two components, one in the direction Ox and the other in the perpendicular direction Oy. The components in direction Ox are equivalent to a single vector

$$\mathbf{X} = \mathbf{P_1} \cos \alpha_1 + \mathbf{P_2} \cos \alpha_2 + ... + \mathbf{P_n} \cos \alpha_n = \Sigma (\mathbf{P} \cos a),$$

and the components in direction Oy are equivalent to a single vector

$$\mathbf{Y} = \mathbf{P}_1 \sin \alpha_1 + \mathbf{P}_2 \sin \alpha_2 + \dots + \mathbf{P}_n \sin \alpha_n = \Sigma \, (\mathbf{P} \sin \alpha).$$

The two vectors $\mathbf{X}$, $\mathbf{Y}$ can now be compounded into a single vector $\mathbf{R}$ making an angle θ with Ox, such that

$$\mathbf{R} \cos \theta = \mathbf{X} \text{ and } \mathbf{R} \sin \theta = \mathbf{Y},$$

and therefore $\quad R^2 = X^2 + Y^2$ and $\tan \theta = Y/X$ ....(1).

2·5. Vectors may be multiplied and divided by scalar numbers. Thus if we take n equal vectors $\overline{AB}$ and compound them together we get a vector $\overline{AC}$, such that $\overline{AC} = n \overline{AB}$; and, conversely, $\overline{AB} = \dfrac{1}{n} \overline{AC}$.

Note that relations of the form $\overline{AC} = n\overline{AB}$, or $p\overline{AB} + q\overline{AC} = 0$ imply that the points A, B, C are in the same straight line.

2·6. Centroids or Mean Centres. If m_1, m_2, $m_3 \dots m_n$ be a set of scalar magnitudes associated with a set of points A_1, A_2, $A_3 \dots A_n$, the centroid or mean centre of the points for the given magnitudes is the point obtained by the following process: Divide the line $A_1 A_2$ at B_1 so that $m_1 A_1 B_1 = m_2 B_1 A_2$; divide $B_1 A_3$ at B_2 so that $(m_1 + m_2) B_1 B_2 = m_3 B_2 A_3$; divide $B_2 A_4$ at B_3 so that $(m_1 + m_2 + m_3) B_2 B_3 = m_4 B_3 A_4$. Proceed in this way until all the points have been connected then the last point of division B_{n-1}, usually denoted by the letter G, is called the *centroid* or *mean centre*.

2·61. In order to shew that this process leads in general to a unique point, i.e. that the point determined by the process is independent of the order in which the points A_1, $A_2 \dots A_n$ are joined, we shall first prove that

$$m_1 \overline{A_1 G} + m_2 \overline{A_2 G} + \dots$$
$$+ m_n \overline{A_n G} = 0 \ \dots(1).$$

Assume that this formula is true for the first r points, i.e. that

$$m_1 \overline{A_1 B_{r-1}} + m_2 \overline{A_2 B_{r-1}} + \dots + m_r \overline{A_r B_{r-1}} = 0.$$

Now the next step in the process is to divide $B_{r-1}A_{r+1}$ at B_r so that

$$(m_1 + m_2 + \ldots + m_r)\overline{B_{r-1}B_r} = m_{r+1}\overline{B_rA_{r+1}},$$

therefore by adding the last two lines

$$m_1\overline{A_1B_r} + m_2\overline{A_2B_r} + \ldots + m_r\overline{A_rB_r} + m_{r+1}\overline{A_{r+1}B_r} = 0.$$

It follows that if the formula (1) is true for r points it is also true for $r + 1$; but it is true for two points, since, by hypothesis, $m_1\overline{A_1B_1} + m_2\overline{A_2B_1} = 0$. Therefore the formula (1) is true for any number of points.

Now if by taking the points in a different order we arrive at a centroid G' we can shew similarly that

$$m_1\overline{A_1G'} + m_2\overline{A_2G'} + \ldots + m_n\overline{A_nG'} = 0 \ldots\ldots\ldots(2);$$

and by subtracting (1) from (2) we get

$$(m_1 + m_2 + \ldots + m_n)\overline{GG'} = 0.$$

Hence G' must coincide with G unless $m_1 + m_2 + \ldots + m_n = 0$. In the latter case there is no centroid at a finite distance, because the last step in the process of finding the centroid consists in dividing a line in the ratio $m_1 + m_2 + \ldots + m_{n-1} : m_n$, i.e. in the ratio $1 : -1$.

2·7. Centroid Method of Compounding Vectors. To shew, with the notation of the last Article, that, if O be any other point, the resultant of n vectors $m_1\overline{OA_1}$, $m_2\overline{OA_2} \ldots m_n\overline{OA_n}$ is $(m_1 + m_2 + \ldots + m_n)\overline{OG}$, where G is the centroid of the points $A_1, A_2 \ldots A_n$ for the magnitudes $m_1, m_2 \ldots m_n$.

This follows at once by substituting

$$\overline{OA_1} = \overline{OG} + \overline{GA_1}, \ \overline{OA_2} = \overline{OG} + \overline{GA_2}, \text{ etc.,}$$

so that

$$m_1\overline{OA_1} + m_2\overline{OA_2} + \ldots + m_n\overline{OA_n}$$

$$= (m_1 + m_2 + \ldots + m_n)\overline{OG} + (m_1\overline{GA_1} + m_2\overline{GA_2} + \ldots + m_n\overline{GA_n});$$

and by **2·61** (1) the sum of the terms in the last bracket is zero, therefore

$$m_1\overline{OA_1} + m_2\overline{OA_2} + \ldots + m_n\overline{OA_n} = (m_1 + m_2 + \ldots + m_n)\overline{OG}.$$

2·8. When reference is made to the centroid of a set of points without mention of any associated magnitudes it is understood that the magnitudes are equal; thus the centroid of a triangle ABC is a point G such that

$$\overline{AG} + \overline{BG} + \overline{CG} = 0.$$

2·9. It may be noticed that if **P**, **Q**, **R** are vectors in the lines OA, OB, OC then the resultant vector is

$$\left(\frac{P}{OA} + \frac{Q}{OB} + \frac{R}{OC} \right) \overline{OG},$$

where G is the centroid of the points A, B, C for the magnitude P/OA, Q/OB, R/OC; for a vector **P** is the same as $\dfrac{P}{OA} \overline{OA}$.

EXAMPLES

1. Prove that, if G is the middle point of AB and G' is the middle point of $A'B'$, then $\overline{AA'} + \overline{BB'} = 2\overline{GG'}$.

2. Prove that, if G is the centroid of n points A_1, $A_2 \dots A_n$, and G' is the centroid of n points B_1, $B_2 \dots B_n$, then

$$\overline{A_1 B_1} + \overline{A_2 B_2} + \dots + \overline{A_n B_n} = n\overline{GG'}.$$

3. Prove that, if H is the orthocentre and O is the circumcentre of a triangle ABC, then

$$\overline{AH} \tan A + \overline{BH} \tan B + \overline{CH} \tan C = 0,$$

and $\qquad\qquad \overline{AO} \sin 2A + \overline{BO} \sin 2B + \overline{CO} \sin 2C = 0.$

4. Shew that, if $m\overline{OA} + n\overline{OB} + p\overline{OC} = 0$ and $m + n + p = 0$, then the points A, B, C are collinear.

Chapter III

RECTILINEAR MOTION. KINEMATICS

3·1. Consider the motion of a point along a straight line or axis Ox on which O is a fixed point. Let P, P' denote the positions of the moving point at times t, t'; and let $OP = x, OP' = x'$. The displacement of the point in time $t' - t$ is $x' - x$, and $(x' - x)/(t' - t)$ is the average rate of displacement or the **average velocity** during the interval $t' - t$. If this ratio be independent of the interval $t' - t$; i.e. if it has the same value for all intervals of time, then the velocity is constant or *uniform*, and equal distances will be traversed in equal times.

Whether the ratio $(x' - x)/(t' - t)$ be constant or not, its limiting value as t' tends to t is defined to be the **measure of the velocity** of the moving point at time t. But this limiting value is the differential coefficient of x with regard to t, so that if we denote the velocity by v, we have

$$v = dx/dt.$$

If on squared paper we plot a curve in which abscissae represent time and ordinates represent distances traversed, the curve is called the *space-time curve*. The curve gives a graphical representation of the motion, because it exhibits graphically the relation between the time and the distance traversed in

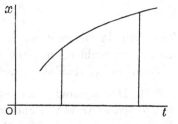

that time. Also the gradient of the curve, i.e. the tangent of the angle that the tangent to the curve makes with the time-axis, gives the value of the velocity dx/dt.

3·2. Acceleration is similarly defined as the rate of change of velocity. Thus, if v, v' denote the velocities of the moving point at times t, t', then $v' - v$ is the change of velocity in time

$t' - t$, and $(v' - v)/(t' - t)$ is the average rate of change of velocity during the interval $t' - t$. If this ratio is independent of the interval $t' - t$, then the acceleration is constant or *uniform*, or equal increments of velocity take place in equal intervals.

Whether the ratio $(v' - v)/(t' - t)$ be constant or not, its limiting value as t' tends to t is defined to be the **measure of the acceleration** of the moving point at time t. But this limiting value is the differential coefficient of v with regard to t, so that if we denote the acceleration by f, we have

$$f = dv/dt = d^2x/dt^2.$$

Following Newton, it is usual to denote differential coefficients with regard to time by dots; thus $\dot{x}$ means dx/dt, and $\ddot{x}$ means d^2x/dt^2.

If on squared paper we plot a curve in which abscissae represent time and ordinates represent velocity, the curve is called the *velocity-time curve*. The gradient of the curve gives the acceleration dv/dt at any instant.

Also the area under the velocity-time curve

$$= \int v\,dt = \int \frac{dx}{dt}\,dt = \int dx = [x]$$

taken between proper limits

= the distance covered in the corresponding time.

3·3. Acceleration represented as a space rate of change. Since $v = dx/dt$ and $f = dv/dt$,

therefore $$f = \frac{dv}{dx}\frac{dx}{dt} = v\frac{dv}{dx}.$$

This formula for acceleration is very important, as it has to be used in all problems in which the velocity is given in different positions rather than at different times.

3·31. If on squared paper a curve be plotted in which abscissae denote spaces traversed and ordinates represent velocities, the curve is called a *velocity-space curve*. The gradient of the curve is dv/dx, and in the figure, in which PN is ordinate v and PG is normal at P, we have

$$\tan NPG = dv/dx,$$

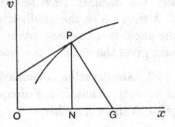

so that   the subnormal $NG = v\,dv/dx$

= the acceleration.

Note that if the acceleration is constant the velocity-space curve is a parabola, since this is the only curve for which the subnormal is constant.

3·4. Units. Measurements of physical quantities must necessarily be expressed as multiples or submultiples of certain units. The choice of the units of length and time is arbitrary; thus the unit of *length* may be a *foot* or a *centimetre*, and the unit of time may be the *mean solar second* or it may be the *sidereal day* (**1·3**), the former being the more convenient unit for purposes outside an observatory as it is mean solar time that is recorded by ordinary chronometers, watches and clocks.

The units of velocity and acceleration are derived units in that they depend on the units of length and time, which may be regarded as fundamental units. The *unit of velocity* is a unit of length described in a unit of time. The *unit of acceleration* is a unit of velocity added in a unit of time. The unit of velocity therefore varies directly as the unit of length and inversely as the unit of time, and the unit of acceleration therefore varies directly as the unit of length and inversely as the *square* of the unit of time. These facts may also be expressed by saying that the unit of velocity is of one dimension in length and minus one dimension in time, and that the unit of acceleration is of one dimension in length and minus two dimensions in time; or, if **L, T** denote the units of length and time, the unit of velocity is $\mathbf{LT^{-1}}$ and the unit of acceleration is $\mathbf{LT^{-2}}$.

The measure of any given quantity varies inversely as the unit chosen; e.g. a velocity measured in yards per second is one-third of the same velocity measured in feet per second.

3·41. Change of Units. If v denote the measure of a velocity and f the measure of an acceleration when **L, T** are the units of length and time, and v', f' denote the measures of the same velocity and the same acceleration when **L′, T′** are the units of length and time, then it is clear that

$$v\mathbf{LT^{-1}} = v'\mathbf{L'T'^{-1}}$$

and $$f\mathbf{LT^{-2}} = f'\mathbf{L'T'^{-2}},$$

since the expressions equated are equivalent representations of
the same velocity and of the same acceleration.

For example, if $\mathbf{L} = 1$ foot, and $\mathbf{T} = 1$ second
and $\mathbf{L}' = 1$ mile and $\mathbf{T}' = 1$ hour,
then $\mathbf{L}' = 5280\mathbf{L}$ and $\mathbf{T}' = 3600\mathbf{T}$,
and, if v' is a velocity of 60 miles per hour,

$$v = 60 \times 5280 \div 3600$$
$$= 88 \text{ ft. per sec.}$$

3·42. Units in Graphical Work. In using the graphical
methods indicated in **3·1, 3·2** and **3·3** attention must be paid to
scales of measurement.

Space-time curve.

If one inch in the abscissa t represents a seconds and one inch
in the ordinate x represents b feet, then the measure of the
velocity is $\dfrac{b}{a}\dfrac{dx}{dt}$ ft. per sec. where dx/dt is the actual gradient in
the diagram.

Velocity-time curve.

If one inch in the abscissa t represents a seconds and one inch
in the ordinate v represents a velocity of b ft. per sec., then the
measure of the acceleration is $\dfrac{b}{a}\dfrac{dv}{dt}$ ft. per sec. per sec. where
dv/dt is the actual gradient in the diagram; and the area under
the curve represents distance covered on the scale

one square inch $= ab$ feet.

Velocity-space curve.

If one inch in the abscissa x represents a feet and one inch in
the ordinate v represents a velocity of b ft. per sec., then the
measure of the acceleration is $\dfrac{b^2}{a}\dfrac{v\,dv}{dx}$ where $v\,dv/dx$ is the actual
subnormal in inches; i.e. the subnormal gives the acceleration
on the scale

1 inch $= b^2/a$ ft. per sec. per sec.

3·5. Uniformly Accelerated Motion. In uniformly accelerated motion the acceleration f is constant, so that by integrating
the relation
$$dv/dt = f$$
we get $v = ft + C,$

where C is constant, and if the velocity at time $t = 0$ be u (called the initial velocity), we must have $C = u$, and therefore

$$v = u + ft \quad \dots\dots\dots\dots\dots\dots(1);$$

though it must be remarked that this relation is otherwise obvious, because, when the acceleration f is constant, the velocity is increased by f in each unit of time.

Again, since $\qquad dx/dt = v = u + ft,$
therefore by integration

$$x = ut + \tfrac{1}{2}ft^2 + C',$$

where C' is constant, and if the origin from which x is measured is taken at the position of the moving point when $t = 0$, we have $x = 0$ when $t = 0$, so that $C' = 0$ and

$$x = ut + \tfrac{1}{2}ft^2 \quad \dots\dots\dots\dots\dots\dots(2).$$

If we now eliminate t between (1) and (2) we get

$$v^2 = u^2 + 2fx \quad \dots\dots\dots\dots\dots\dots(3).$$

This last result can also be obtained by integrating the alternative expression for acceleration, $v\,dv/dx = f$, which, when f is constant, gives
$$v^2 = 2fx + C'',$$

where C'' is constant; and since $v = u$ when $x = 0$, therefore $C'' = u^2$ and
$$v^2 = u^2 + 2fx.$$

The student will do well to note that the formulae of this article are only true when the acceleration is constant.

3·51. Formulae (1) and (2) of the last Article follow graphically from the velocity-time curve. Since the gradient of the curve represents the acceleration and this is constant, therefore the curve in this case is a straight line inclined at an angle $\tan^{-1}f$ to the t axis.

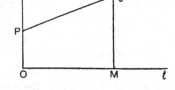

Let PQ be this line and Q the point (t, v), P the point $(0, u)$, where $OM = t$, $MQ = v$, $OP = u$. Then

$\qquad (MQ - OP)/OM = \text{gradient} = f,$
therefore $v = u + ft.$

Again space described in time $t = $ area under curve
$$= OMQP$$
$$= \tfrac{1}{2}OM(OP + MQ)$$
$$= \tfrac{1}{2}t(2u + ft)$$
$$= ut + \tfrac{1}{2}ft^2.$$

3·51. Acceleration due to Gravity. It is an experimental fact, first observed by Galileo, that all bodies which are allowed to fall freely at a given place possess the same constant acceleration, usually denoted by g.

The value of g varies slightly, increasing with the latitude of the place; but the difference between the values at the equator and the poles amounts only to about one-half per cent. The value in the latitude of London is approximately 981 cm. per sec. per sec. or 32·2 ft. per sec. per sec.

3·6. Applications. The law of acceleration in a particular problem may be given by expressing the acceleration as a function of (i) the time t, or (ii) the distance x, or (iii) the velocity v. The problem of further investigating the motion can then be solved as follows :

(i) *Acceleration a given function of the time, say* $\phi(t)$.

We have $\qquad\qquad \dot{v} = \phi(t)$

therefore, by integration, $\quad v = \int\phi(t)\,dt + C,$

or $\qquad\qquad\qquad \dot{x} = \psi(t) + C$, say,

where $\psi(t)$ is the integral of $\phi(t)$. Then another integration gives

$$x = \int\psi(t)\,dt + Ct + C'$$

where the constants C, C' can be determined if the velocity and position at a given time are known.

(ii) *Acceleration a given function of the distance, say* $\phi(x)$.

In this case, there are two ways of proceeding :

(a) We use the form $v\,dv/dx$ for acceleration and write

$$v\,dv/dx = \phi(x),$$

so that, by integration, $\qquad v^2 = 2\int\phi(x)\,dx + C.$

Then, if we put $\psi(x)$ for $2\int\phi(x)\,dx$ and $\dot{x}$ for v,

we have $\qquad\qquad \dot{x} = \pm\{\psi(x) + C\}^{\frac{1}{2}}$

and therefore $\qquad \pm\int\dfrac{dx}{\{\psi(x) + C\}^{\frac{1}{2}}} = t + C'.$

(β) Alternatively, we write

$$\ddot{x} = \phi(x)$$

and multiply both sides of the equation by $2\dot{x}$, so that

$$2\dot{x}\ddot{x} = 2\phi(x)\dot{x}.$$

Then integrate with regard to t and we get

$$\dot{x}^2 = 2\int\phi(x)\,dx + C,$$

and from thence we proceed as in (a) above.

This method of procedure, in which we multiply both sides of the equation by $2\dot{x}$ in order to make it integrable, is of frequent application and deserves special notice.

As an example let the acceleration be $-n^2x$ where, as usual, x denotes distance from a fixed origin and suppose that at time $t=0$ we have $x=a$ and $\dot{x}=0$; i.e. a point starts from rest at a distance a from the origin with acceleration n^2x towards the origin. We put

$$\ddot{x} = -n^2x,$$

then multiply by $2\dot{x}$ and we get

$$2\dot{x}\ddot{x} = -2n^2x\dot{x},$$

which gives on integration

$$\dot{x}^2 = -n^2x^2 + C.$$

But $\dot{x}=0$ when $x=a$, therefore $C=n^2a^2$, and

$$\dot{x}^2 = n^2(a^2 - x^2),$$

or

$$-\frac{dx}{\sqrt{(a^2-x^2)}} = n\,dt;$$

therefore

$$\cos^{-1}\frac{x}{a} = nt + C',$$

and, since $x=a$ when $t=0$, therefore $C'=0$, and

$$x = a\cos nt.$$

Note that the choice of the minus sign on taking the square root is determined by the fact that the point starts from rest with an acceleration directed towards the origin, so that x decreases as t increases, or dx and dt have opposite signs.

In this particular example we might also obtain the same solution by proceeding as in **1·7** (16), where the solution is obtained in the form

$$x = C\cos(nt + \alpha).$$

If we proceed in this way, we have to determine the constants C and α from the conditions that when $t=0$

$$x=a \quad\text{or}\quad C\cos\alpha = a,$$

and

$$\dot{x}=0 \quad\text{or}\quad -nC\sin\alpha = 0.$$

These give $\alpha=0$ and $C=a$, making $x=a\cos nt$ as before.

(iii) *Acceleration a given function of velocity, say* $\phi(v)$.

In this case we may either connect velocity with time by writing

$$dv/dt = \phi(v),$$

and integrating in the form

$$\int\frac{dv}{\phi(v)} = t + C;$$

or we may connect velocity with distance by writing

$$v\,dv/dx = \phi(v),$$

and integrating in the form

$$\int\frac{v\,dv}{\phi(v)} = x + C.$$

Whether it is then possible to proceed to a further integration after putting $\dot{x}$ for v will depend on the form of the functions.

Example. A particle moves in a straight line under a retardation kv^{m+1}, where v is the velocity at time t. Shew that, if u is the velocity when $t=0$, then

$$kt = \frac{1}{m}\left(\frac{1}{v^m} - \frac{1}{u^m}\right),$$

and find a corresponding formula for space in terms of v.

We have
$$\dot{v} = -kv^{m+1};$$

therefore
$$\int \frac{dv}{v^{m+1}} + C = -\int k\,dt,$$

or
$$\frac{1}{mv^m} - C = kt,$$

but $v = u$ when $t = 0$, therefore $C = 1/mu^m$

and
$$kt = \frac{1}{m}\left(\frac{1}{v^m} - \frac{1}{u^m}\right).$$

Again
$$v\frac{dv}{dx} = -kv^{m+1},$$

therefore
$$\int \frac{dv}{v^m} + C = -\int k\,dx$$

or
$$\frac{1}{(m-1)v^{m-1}} - C = kx,$$

and, if x be measured from the position in which $v = u$, we have $x = 0$ when $v = u$, therefore $C = 1/(m-1)u^{m-1}$ and

$$kx = \frac{1}{m-1}\left(\frac{1}{v^{m-1}} - \frac{1}{u^{m-1}}\right).$$

3·7. Graphical Methods. If a table of corresponding values of any two of the four magnitudes x, t, v, f is given, it is in general possible to determine the other two by plotting suitable graphs, based upon the relations $v = dx/dt,\ f = dv/dt = v\,dv/dx$.

For example, suppose a table of corresponding values of v and f to be given. Since $f = dv/dt$,

therefore
$$dt = \frac{dv}{f}, \text{ or } t = \int\frac{dv}{f}.$$

Hence if we plot a curve in which the abscissae denote v and the ordinates denote $1/f$, then the area under the curve up to any point will represent the time t.

The corresponding values of the distance x can then be found from $v\,dv/dx = f$, which gives

$$dx = \frac{v\,dv}{f} \text{ and therefore } x = \int\frac{v}{f}\,dv.$$

So that if we plot a curve in which the abscissae denote v and the ordinates denote v/f, then the area under the curve up to any point will represent the distance x.

Alternatively, x can be found by using the formula $v = dx/dt$, which gives $x = \int v\,dt$, if we first tabulate the values of t found above and then plot a velocity-time curve. The area under the curve up to any point will represent the distance x.

3·8. Examples. (i) *A train passes a station A at* 40 *miles per hour and maintains this speed for* 7 *miles, and is then uniformly retarded, stopping at B which is* 8 *miles from A. A second train starts from A the instant the first train passes and being uniformly accelerated for part of the journey and uniformly retarded for the rest stops at B at the same time as the first train. What is its greatest speed on the journey?*   [S. 1910]

This example illustrates the utility of the velocity-time curve. The units chosen are a mile per hour for v, and an hour for t. For the first train the velocity-time curve consists of the straight line PQ or $v = 40$, and the straight line QR along which the velocity is uniformly retarded to zero.

OM, MR are the times required for the first seven miles and the eighth mile respectively; and the areas under the curve, i.e. $OMQP$ and MRQ represent 7 miles and 1 mile respectively.

Therefore $40\,OM = 7$ and $20\,MR = 1$.

Again, for the second train which is uniformly accelerated and then uniformly retarded, starting from rest and ending at rest and covering the same distance in the same time, the velocity-time curve consists of two straight lines OS, SR with the condition that the area OSR represents 8 miles. Hence the greatest speed SN is given by
$$\tfrac{1}{2}SN\,.\,OR = 8 \; ; \text{ but } OR = \tfrac{9}{40},$$
therefore $SN = 71\tfrac{1}{9}$ miles per hour.

3·81. In the graphical solution of examples it is sometimes useful to remember that if the acceleration increase or decrease steadily with the time then the velocity-time curve is a parabola. For if
$$\dot{v} = at + b,$$
where a and b are constants, then
$$v = \tfrac{1}{2}at^2 + bt + c,$$
which represents a parabola if v, t are used as coordinates.

(ii) *A train starts from a station A with acceleration ·8 ft.-sec.⁻², and this acceleration decreases uniformly for 2 minutes, at the end of which the train has acquired full speed which is maintained for another two minutes. Then brakes are applied and produce a constant retardation 3·5 ft.-sec.⁻² and bring the train to rest at B. Draw an acceleration-time curve and a velocity-time curve. Find the maximum velocity and the distance AB.*

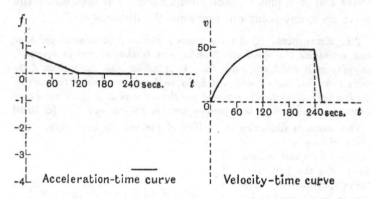

Acceleration-time curve   Velocity-time curve

The *acceleration-time* curve consists of portions of three straight lines as shewn in the first diagram.

The velocity at any time during the first 2 minutes is measured by the corresponding area under the acceleration-time curve :

e.g. velocity at one minute = area of trapezium of base 60 and mean height ·6 = 36 f.s.

and velocity at two minutes = area of triangle of base 120 and height ·8 = 48 f.s.

In this way any number of points on the velocity-time curve during the first two minutes may be plotted. At two minutes the velocity-time curve becomes a straight line parallel to the time axis. At four minutes there is a uniform retardation 3·5 f.s.s which will destroy the velocity of 48 f.s. in 48/3·5 secs. or 13·7 secs., so that the last part of the velocity-time curve is another straight line meeting the time axis at $t = 253·7$ secs.

The total distance travelled is given by the area under the velocity-time curve. The area under the curved portion can be estimated by counting squares or by dividing it into parallel strips and using mean ordinates; or, more accurately from the fact that the curve is a parabola with axis vertical and vertex at (120, 48), so that the area is ⅔ of the rectangle whose sides are 120 and 48 = 3840. The remaining area is

$$\text{a rectangle } 48 \times 120 = 5760,$$

$$\text{and a triangle } \tfrac{1}{2} \times 48 \times 13·7 = 329,$$

making the total distance traversed 9929 ft.

This problem may also be solved by direct analysis. Thus, since the acceleration falls steadily from ·8 f.s.s to zero in 120 secs. therefore

$$\text{acceleration after } t \text{ secs.} = \cdot 8 \left(1 - \frac{t}{120}\right),$$

or

$$\ddot{x} = \cdot 8 \left(1 - \frac{t}{120}\right) \text{ ft.-sec.}^{-2},$$

so that

$$\dot{x} = \cdot 8 \left(t - \frac{t^2}{240}\right) \text{ ft.-sec.}^{-1},$$

and

$$x = \cdot 8 \left(\frac{t^2}{2} - \frac{t^3}{720}\right) \text{ ft.};$$

no constants of integration being required since $\dot{x} = 0$ and $x = 0$ when $t = 0$.

It follows that when $t = 120$ we have $\dot{x} = 48$ ft./sec. and $x = 3840$ and the rest of the solution is as above.

EXAMPLES

Uniform Velocity

1. Two trains take 3 seconds to pass one another when going in opposite directions, but only 2·5 seconds if the speed of one is increased by 50 per cent. How long would one take to pass the other when going in the same direction at their original speeds ?

2. A steamer takes m minutes to go a mile downstream and n minutes to go a mile upstream. Find the speed of the current and of the steamer relative to it.

3. A line of men are running along a road at 8 miles an hour behind one another at equal intervals of 20 yards. A line of cyclists are riding in the same direction at 15 miles an hour at equal intervals of 30 yards. At what speed must an observer travel along the road so that whenever he meets a runner he also meets a cyclist ?

Uniform Acceleration

4. A point moving with uniform acceleration describes distances s_1, s_2 feet in successive intervals of t_1, t_2 seconds. Prove that the acceleration is

$$2\left(s_2 t_1 - s_1 t_2\right)/t_1 t_2 \left(t_1 + t_2\right).$$

5. A point moves with uniform acceleration and v_1, v_2, v_3 denote the average velocities in three successive intervals of time t_1, t_2, t_3. Prove that

$$v_1 - v_2 : v_2 - v_3 = t_1 + t_2 : t_2 + t_3.$$

6. A heavy particle is projected vertically upwards. Shew that if t_1, t_2 are the times at which it passes a point at a height h above the point of projection in ascending and descending, then $t_1 t_2 = 2h/g$.

7. A point moves from rest with uniform acceleration. Shew that in any interval the space-average of the velocity is 4/3 of the time-average.

8. An express train is sent off in two parts at an interval of 5 minutes. If both parts are uniformly accelerated and attain their maximum speed of 60 miles per hour in a mile, prove that the first part has gone 4 miles before the second starts and that at full speed they run 6 miles apart.

9. A body moving in a straight line travels distances AB, BC, CD of 153 feet, 320 feet, 135 feet respectively in three successive intervals of 3 secs., 8 secs., and 5 secs. Shew that these facts are consistent with the hypothesis that the body is subjected to uniform retardation. On this hypothesis find the distance from D to the point where the velocity vanishes, and the time occupied in describing this distance. [M. T. 1915]

10. A train which starts from rest and accelerates in 1 mile to a full speed of 40 m.p.h. and stops in $\frac{1}{2}$ a mile, has, in the course of a journey, to be slowed down to 20 m.p.h. for a distance of 2 miles. Shew that in consequence it arrives at its destination $3\frac{9}{16}$ mins. late, assuming that acceleration is always at the same uniform rate and that retardation is also always at the same uniform rate.

11. Two stopping points of an electric tramcar are 440 yards apart. The maximum speed of the car is 20 miles per hour and it covers the distance between stops in 75 seconds. If both acceleration and retardation are uniform and the latter is twice as great as the former, find the value of each of them, and also how far the car runs at its maximum speed.

[S. 1924]

Variable Acceleration

12. A body moving in a straight line describes the following distances in the given times :

Time in seconds	0	5	10	15	20	25	30
Distance in feet	5	13·75	60	176·25	400	768·75	1320

Deduce approximately the velocity-time and acceleration-time curves for the same period. [S. 1911]

13. The relation between acceleration and time for a car starting from rest is given by the table

Time in seconds	0	10	20	30	40	50	60	70
Accel. in ft./sec.²	·5	1	1·2	·5	− ·3	− ·45	− ·45	− ·3

Draw the acceleration-time curve and deduce the velocity-time and distance-time curves.

14. The relation between acceleration and distance for a car starting from rest is given by the table

Distance in feet	0	100	200	400	600	800	1000
Accel. in ft./sec.^{2}	2	1·7	1·4	·85	·2	− ·2	0

Draw the acceleration-distance curve and deduce the velocity-distance curve.

15. If the relation between x and t is of the form $t = \alpha x^2 + \beta x$, find the velocity v as a function of x, and prove that the retardation of the particle is $2\alpha v^3$.

The following observations were made on a rifle-bullet:

x	0	150	300	ft.
t	0	·0698	·1408	sec.

Assuming that the relation between x, t is that stated above, find α, β and shew that the velocity at the instant of the second observation was 2131 ft. per sec. approximately.   [S. 1912]

16. By proper choice of units the curve on a time base representing the acceleration of an electric train is a quadrant of a circle, whose centre is the origin. The initial acceleration is 2·5 ft. per sec. per sec., and the acceleration falls to zero in 20 seconds. Calculate the velocity acquired and the distance described in that time.   [S. 1917]

17. A particle moving in a straight line is subject to a resistance which produces the retardation kv^3, where v is the velocity and k is a constant. Shew that v and t (the time) are given in terms of s (the distance) by the equations
$$v = u/(1 + ksu), \quad t = \tfrac{1}{2} ks^2 + s/u,$$
where u is the initial velocity.

As a result of certain experiments with a rifle, it was estimated that the bullet left the muzzle with a velocity of 2400 ft. per sec. and that the velocity was reduced to 2350 ft. per sec. when 100 yards had been traversed. Assuming that the air-resistance varied as v^3, and neglecting gravity, calculate the time of traversing 1000 yds.   [M. T. 1913]

ANSWERS

1. 15 secs.   2. $30(n-m)/mn$, $30(n+m)/mn$ miles per hour.
3. 6 m.p.h.   9. 121 ft., 11 secs.   11. $1\frac{11}{15}$ f.s.s; $\frac{22}{15}$ f.s.s; 440 ft.
16. 39·27 f.s.; 452 ft.   17. 1·38 secs.

Chapter IV

RECTILINEAR MOTION. KINETICS

4·1. In the last chapter we discussed the measurement of the velocity and acceleration of a point moving in a straight line. We must now begin to study the branch of Dynamics called Kinetics, which is concerned with the effects of forces on the motion of bodies.

The definitions and laws of motion enunciated by Newton in the seventeenth century form the foundation of a science of dynamics which has been developed by many other mathematicians. This dynamics, now commonly called 'Newtonian Mechanics,' is the basis of all the theoretical work in applied mechanics or engineering and the results of the theory have been and are still being confirmed every day by numerous appeals to experiment; so that nowadays no one questions whether the theory is adequate to furnish reliable results in common matters. It is only when on the one hand we begin to investigate what is relatively very small, e.g. the interior of an atom, or on the other hand when we leave the Earth and apply the Newtonian theory on a much wider scale that discrepancies can be detected, and even here it may be remarked that Newtonian Mechanics has proved adequate to enable astronomers to predict the time of eclipses with an accuracy that would hardly have been possible had the foundations of dynamics been radically at fault.

From a philosophical standpoint Newton's definitions and laws of motion offer much scope for criticism, particularly his assumptions of absolute time and space. But it is no part of our purpose to discuss this aspect of the matter; we intend to leave to the reader his primary conceptions of time and space and adopt a Newtonian basis for the development of the subject.

4·11. Force. It is a fact of everyday experience that bodies move more often in curved paths and with varying speed than in straight paths at uniform speed. We assume that it is the action of other bodies that cause the speed of a body to vary or its path to bend. We describe this process by saying that the bodies are exerting forces on one another, and we define *force* as that which changes or tends to change the state of motion of a body.

Newton's First Law of Motion that *Every body perseveres in its state of rest or of moving uniformly in a straight line, except in so far as it is made to change that state by external forces* is implied in the foregoing assumption and definition of force. This law is also commonly called the *Law of Inertia*.

4·12. Mass. Matter is one of the primary conceptions of the mind, it cannot be defined satisfactorily, but it is clear that any definition of matter would have to embrace all things that can be perceived by the sense of touch and some things which are perceived by other senses. The mode by which the presence of matter is most easily perceived is through the effort required to produce in it a sudden change of motion. This property possessed by matter is called *inertia*. The measure of the inertia of a body is called its *mass*.

The familiar process of 'weighing' on a common balance is the process by which masses are compared. A scale of comparison is instituted in which we fix upon a certain body A as a unit and determine that another body B is equivalent, in the sense of balancing, to some multiple or sub-multiple of the unit A. In this way every body has associated with it a number called its *mass*.

4·13. A Material Particle is defined to be a body so small that, *for the purposes of our investigation*, the distances between its different parts may be neglected*.

4·14. The Momentum of a material particle is the product of its mass and velocity.

4·15. Measurement of Force. Newton's Second Law of Motion states that *Change of motion is proportional to the impressed force and takes place in the direction in which the force is impressed.* In modern phraseology we substitute the words *rate of change of momentum* for *change of motion*,

* Clerk Maxwell, *Matter and Motion*, Art. VI.

and we see that the law is an assertion of the mode of measurement of force. It is proportional to rate of change of momentum. Hence in the rectilinear motion of a particle of mass m, if v be its velocity and f its acceleration it is acted on by a force in the direction of motion proportional to $\frac{d}{dt}(mv)$ or mf. And if we take the unit of force to be the force which acting upon a particle of unit mass causes it to have unit acceleration we may write the force $P = mf$. The acceleration lasts so long as the force acts and ceases when the force ceases to act.

4·16. Force as a Vector. The second law of motion implies much more than is asserted in the last article. It implies that if a body A is in motion in the presence of several other bodies B, C, etc. which exert forces P, Q, etc. upon it, either in the same or in different directions, then each force produces its own contribution to the acceleration of the body A, and this contribution is the same in magnitude and direction as it would be if the force considered were the only force acting upon A. This implication of the law is usually called the *Principle of the Physical Independence of Forces.*

Further, if the body A under observation be a material particle whose velocity is accelerated, at any instant its acceleration must possess a definite magnitude and direction and therefore be such as would be produced by a single force of the proper magnitude in the assigned direction. It follows that, if such a particle be acted upon by several forces, P, Q, etc., simultaneously, the combined effect of these forces is the same as that of a single force. That is to say, the separate forces P, Q, etc. are components of a single resultant force and the accelerations they severally produce are components of a single resultant acceleration. And we assert as a **fundamental axiom** that *forces are compounded by the vector law of addition.* The structure of dynamical theory is built upon this axiom, and the justification for the hypothesis is not to be found in any attempts at formal proof but in the general agreement of the theory with practical applications as the results of everyday experience and observation.

4·161. We will now change the order of the steps in the line of argument and state it concisely as follows:

(i) we assert the fact that forces obey the vector law of addition;

(ii) we assert the second law of motion with the implication that when a particle is acted upon by several forces simultaneously its rate of change of momentum is proportional to the resultant of the forces compounded vectorially;

(iii) we also ascribe the vector property to acceleration, and assert the exact equivalence of the force vector and the vector 'mass × acceleration,' so that, when resolved in any assigned directions, the corresponding components of the two vectors are also equal.

4·17. Weight. The weight of a body is the force with which the Earth attracts it. It has already been remarked (**3·51**) that bodies fall to the Earth with a constant acceleration g. It follows that, if W is the weight of a body of mass m, then

$$W = mg.$$

The unit of force to which reference was made in **4·15** is called the absolute unit of force. In the British absolute system of units the unit of mass is called a *pound*. It is the mass of a certain piece of platinum deposited in the office of the Exchequer and defined by Act of Parliament as 'the Imperial Standard Pound Avoirdupois.' When pound, foot, second units are employed the unit of force is called the **poundal**. This is the absolute unit of force in the pound, foot, second system. It is the force which when acting on a body of mass one pound gives it an acceleration of one foot per second per second.

We observe that in accordance with the formula $W = mg$, the weight of a body of mass m lb. is mg poundals, and the weight of a body of mass 1 lb. is g poundals, so that a poundal is equal to a little less than the weight of half an ounce.

There is another system of units, in use among engineers, in which the unit of force is the weight of one pound, the unit of mass being the mass of g lb., and the mass of W lb. weight being W/g units of mass. The argument at the foundation of this system of units is that if different forces act in turn upon the same body the forces are proportional to the accelerations they produce in the body. Thus if a force equal to the weight of P pounds produces an acceleration f ft.-sec.$^{-2}$ in a body of weight W pounds then

$$\frac{P}{W} = \frac{f}{g}.$$

It follows that $\qquad P = \frac{W}{g} f,$

where W is the weight of the body and P the force acting upon it both measured in terms of the same unit, which may clearly be the weight of one pound, or the weight of one ton or the weight of any other mass that may be found convenient.

4·18. c.g.s. units. In France the standard of mass is a piece of platinum called a kilogramme, the thousandth part of which, or gramme, is the unit of mass, the centimetre being the unit of length and the second the unit of time. In this c.g.s. system the absolute unit of force is called a **dyne**. It is the force which acting on a mass of one gramme gives it an acceleration of one centimetre per second per second. Since $g = 981$ cm.-sec.$^{-2}$, therefore the weight of one gramme is 981 dynes.

4·19. It is to be noted that the relation

$$\text{force} = \text{mass} \times \text{acceleration}$$

does not require that the force and acceleration shall be constant. It is a relation that holds good at every instant during motion and may be expressed in the form

$$X = m\ddot{x},$$

where X is the force tending to increase $\dot{x}$.

4·2. Impulse. The *impulse of a constant force* during a given interval of time is defined to be the product of the force and the time during which it acts; i.e. Pt if P is the constant force and t the interval.

The *impulse of a variable force* during a given interval of time is defined to be the time integral of the force for that interval; i.e. $\int P\,dt$ integrated through the given interval.

In every case the

impulse of a force = the change of momentum produced by it.

Thus, using the formulae for constant force and acceleration, the impulse of the force $= Pt$

$$= mft$$

$$= m\,(v - u)$$

$$= \text{change of momentum.}$$

Again, for a variable force $X = m\ddot{x}$,

the impulse of the force $= \displaystyle\int_{t_1}^{t_2} X\,dt$

$$= \int_{t_1}^{t_2} m\ddot{x}\,dt$$

$$= m\dot{x}_2 - m\dot{x}_1,$$

where $\dot{x}_1$, $\dot{x}_2$ are the velocities at the beginning and end of the interval; and, as before, this is the change of momentum produced.

4·21. Force-time curve. If we plot a curve in which abscissae represent time and ordinates represent force the area under the curve will represent the impulse of the force, or the change of momentum produced in the given interval of time.

4·22. Example. *A water jet issues from a nozzle of 2 square inches section with velocity 60 f.s. and strikes a plane surface placed at right angles to the jet. Find the force exerted on the plane.*

The number of cubic feet that strike the plane per second $= \frac{2}{144} \times 60$, and the mass of a cubic foot of water $= 62·5$ lb.

Therefore the momentum destroyed per second

$$= \tfrac{2}{144} \times 60 \times 62·5 \times 60 = 3125 \text{ absolute units.}$$

But if P is the force exerted in poundals its impulse in one second

$$= P \text{ absolute units,}$$

therefore $P = 3125$ poundals

$$= 97·65 \text{ lb. weight.}$$

4·3. Work. A force is said to do work when its point of application undergoes a displacement in the line of action of the force. Thus, if a constant force P has its point of application advanced through a distance s in its line of action, the force is said to do work Ps. If, further, the force P remaining

constant in magnitude and direction, the displacement of the
point of application A be in
an arbitrary direction AA',
the work done by the force is
measured by the product of
the force and the projection
AM of the displacement upon
the line of action of the force,
i.e. in either figure work done
$= P \cdot AA' \cos \theta$, giving a negative result when θ is an

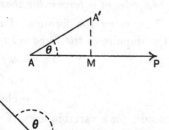

obtuse angle, and no work done when the displacement is perpendicular to the direction of the force P.

More generally, when the point of application of a variable
force is displaced along any curve,
let AA' denote an element ds of
the curve, P the force at A and
θ its inclination to the tangent
to the curve. Regarding the force
as constant during the infinitesimal displacement AA', the work
done in this displacement is
$P \cdot AA' \cos \theta$ or $P \cos \theta \, ds$; we

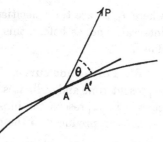

therefore represent the work done in any finite displacement
by an integral
$$\int P \cos \theta \, ds,$$
taken along the curve of displacement from the initial to the
final position.

4·31. Taking one foot as the unit of length and the weight
of one pound as the unit of force, the unit of work is called the
foot-pound. It represents the work that would have to be done
in order to lift one pound vertically through one foot.

Similarly, if one poundal be taken as the unit of force, the
unit of work is called the *foot-poundal*.

4·32. Power is defined as the *rate of doing work*.

The practical unit of power is called the *horse-power* and
represents the doing of 33,000 foot-pounds per minute.

In the absolute C.G.S. system of units, the unit of work is

called the *erg*, and represents the work done when a force of one dyne has its point of application advanced through one centimetre in the direction of the force.

The unit of power in the same system is called the *watt* and represents the doing of 10^7 ergs per second.

4·4. Energy is defined as capacity for doing work. Thus, when a body is in motion, work can be done in overcoming resistance before the body is brought to rest, and this amount of work is called the **Kinetic Energy** of the body.

A body may also possess energy in virtue of its position, represented by the work that would be done by the forces acting on the body if it moved from its stated position to some standard position. This is called **Potential Energy.**

In measuring potential energy it is necessary to choose a standard position in which a body may be considered to possess zero potential energy, and the choice of this standard position is to some extent arbitrary. For example, if we consider the weights of bodies as forces capable of doing work as the bodies descend to a lower level, we may choose to consider bodies on the floor of the room as possessing zero potential energy and then all bodies at a higher level possess a positive amount of potential energy, measured in each case by the weight of the body multiplied by the height above the floor. On the other hand, we might choose the ceiling of the room as the level of zero potential energy, and then all bodies below the ceiling would possess negative potential energy. In a particular problem it is not the absolute value of the potential energy that is important but the *change* of potential energy that takes place in a movement of the body under consideration, and therefore the arbitrariness of the choice of the zero of potential energy simplifies rather than complicates a problem.

4·41. Formula for Kinetic Energy. Suppose that the motion of a body of mass m moving with velocity u is opposed by a constant force P and brought to rest in a distance x. The work done against the resistance $= Px$

$$= mfx,$$

when f is the constant retardation produced by P.

But since the body comes to rest in a distance x, therefore

$$0 = u^2 - 2fx.$$

Hence we have $\qquad Px = \frac{1}{2}mu^2,$

i.e. $\frac{1}{2}mu^2$ is the energy possessed by the body in virtue of its motion; so the Kinetic Energy of a body of mass m moving with velocity u without rotation is defined to be $\frac{1}{2}mu^2$.

4·42. Principle of Work. Conservation of Energy.

In any displacement of a body the change in the Kinetic Energy is equal to the work done by the forces.

This is a general proposition of which a proof will be given in a later chapter. At present we confine ourselves to rectilinear motion, and write the connection between acceleration and force in the form

$$mv\,dv/dx = X\,;$$

therefore $\qquad \int mv\,dv = \int X\,dx + C,$

or, if u, v denote the velocities in the positions $x = x_1$, $x = x_2$, we have

$$\tfrac{1}{2}mv^2 - \tfrac{1}{2}mu^2 = \int_{x_1}^{x_2} X\,dx,$$

which is the required result.

Since 'work done' means an equivalent loss of Potential Energy, the last result implies that in any displacement

gain of Kinetic Energy = loss of Potential Energy,

so that the sum of the Kinetic and Potential Energies is constant throughout the motion. This is the **Principle of Conservation of Energy**. The principle applies to dynamical systems of the most general kind, but we observe that so far we have only proved it for the case of rectilinear motion of a single body.

4·43. Force-space Curve.

If we plot a curve in which ordinates represent the force acting on a body in the direction of its motion, and abscissae represent distances traversed, the area under the curve, e.g. $\int X\,dx$, represents the work done or the increase in the Kinetic Energy in traversing the distance under consideration.

4·44. Efficiency. In most machines some part of the work done is expended in overcoming friction, or resistances of that nature; this portion of the work is regarded as 'wasted' and the remaining portion is described as 'useful work.' The ratio of the useful work to the whole work done is called the **efficiency** of the machine.

4·45. Examples. (i) *A weight of* 200 *lb. hangs freely from the end of a rope. The weight is hauled up vertically from rest by winding up the rope. The pull starts at* 250 *lb. and diminishes uniformly at the rate of* 1 *lb. for every foot wound up. Find the velocity after* 30 *ft. have been wound up, neglecting the weight of the rope.*

The figure shows the force-space diagram, the force decreasing from 250 to 220 pounds weight in a distance 30 ft.

The work done is therefore

$$= \tfrac{1}{2}(250 + 220) \times 30 \text{ ft. lb.}$$
$$= 7050 \text{ ft. lb.}$$

But an amount of work

$$= 200 \times 30 = 6000 \text{ ft. lb.}$$

is done in lifting 200 lb. through 30 ft. Therefore the additional work 1050 ft. lb. is converted into kinetic energy; and if v is the velocity acquired

$$\tfrac{1}{2} \times 200 v^2 = 1050g \text{ foot-poundals,}$$

therefore　　　　　　　　　$v = 18 \cdot 3 \text{ f.s.}$

This problem may also be solved by the consideration that the upward pull after ascending x feet is $(250 - x)$ pounds weight; and therefore, subtracting the weight of the body, the resultant upward force on the body is $(50 - x)g$ poundals.

Hence　　　　　　　$200 v\, dv/dx = (50 - x)g,$

and, by integrating, $100v^2 = (50x - \tfrac{1}{2}x^2)g$; the constant of integration being zero because $v = 0$ when $x = 0$. For $x = 30$ this gives $v = 18 \cdot 3$ f.s. as before.

(ii) *A stream of water,* 1 *square foot in section, flowing at the rate of* 16 *ft. per sec. enters a turbine. At what rate, in horse-power, does the water deliver energy?* (1 *c.f. of water* = 62·5 *lb.*)

What fraction of this energy is used when the water power drives a shaft, at a speed of 100 *revolutions per minute, with a couple of which the moment is* 140 *with the pound weight and foot as units?*　　　　[M. T. 1909]

16 c.f. of water have a mass of 1000 lb.; and in each second this mass of water with velocity 16 f.s. enters the turbine. The kinetic energy of this mass

$$= \tfrac{1}{2} \times 1000 \times 16^2 = 128000 \text{ foot-poundals.}$$

This is the energy delivered per second, and this rate of working is equivalent to $128000/550g$ horse-power

$$= 7\tfrac{3}{11} \text{ H.P.}$$

The work done by a couple in any displacement is the product of the moment of the couple and the angle through which its arm is turned. For since all couples of the same moment in the same plane are equivalent it is immaterial about what point the arm turns. Let the couple, of moment G, be represented by equal forces P at the ends of an arm AB. Let the arm turn through a small angle $d\theta$ about the end A, so that B moves to B'.
The work done $= P \cdot BB'$

$$= P \cdot AB\, d\theta = G \cdot d\theta.$$

And, by summation, the work done for any finite rotation θ is $G\theta$.

Hence in the particular case considered the work done on the shaft $= 140 \times 200\pi$ foot-pounds per minute and this is equivalent to $140 \times 200\pi/33000$ H.P.

$$= 28\pi/33 \text{ H.P.}$$

It follows that the fraction of the energy used is $7\pi/60$ or about $\tfrac{11}{30}$ of the whole, and this fraction represents the efficiency of the machine.

4·5. Applications. Locomotive Engines and Motor Cars.

The propulsive force in the case of a locomotive engine or a motor car is the friction between the driving wheels and the ground. The driving wheels are made to rotate by means of a crank attached to the axle. If there were no friction between the wheels and the ground the car would remain stationary while the wheels slipped round, but the slipping of the wheels on the ground is

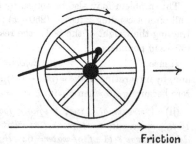

Friction

opposed by friction and this frictional force causes the forward motion of the car. If R denote the pressure of the driving wheels on the ground and μ the coefficient of friction, the propulsive force cannot exceed μR.

In the case of the other wheels of the car or train, the wheels which run freely without compelling cranks, it is the friction

between the wheels and the ground which causes the rotation.
For if there were no friction the
wheels would slide forward without revolving; the friction on
them is therefore in the opposite
direction to the motion of the
car, but its amount is small compared to the frictional force on
the driving wheels being no more
than sufficient to cause the rotation of the wheels.

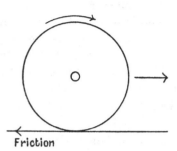

Friction

In order to stop a car or train brakes are applied to the wheels.
This is effected by pressing brake
shoes either on to the rims of the
wheels or on to the rims of smaller
concentric wheels rigidly attached
to the actual running wheels. This
results in frictional forces f being
set up opposing the rotation of
the wheels, and this necessitates a
larger frictional force F between the

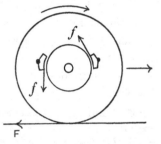

wheels and the ground if the same speed is maintained. But
the friction F opposes the motion of the car as a whole and the
result is a reduction of the speed. The maximum value of F is
μw where w is the load carried by the wheel, hence assuming the
same coefficient of friction at all contacts the pressure of the brake
shoes must not exceed w or the wheel will become locked and
skidding will ensue. The maximum retarding force that brakes
can produce is therefore μ times the weight of the car or train.

4·51. Example. *The weight of a train is* 400 *tons, the part of the
weight of the engine carried by the driving wheels is* 30 *tons, and the
coefficient of friction between the driving wheels and the rails is* ·16. *Shew
that at the end of a minute after starting on the flat the velocity will be less
than* 15·8 *miles per hour.*   [M. T. 1909]

The propulsive force is ·16 × 30 tons weight, so if v ft. per sec. be the
velocity acquired when all frictional resistance to motion is neglected, by
equating the momentum set up to the impulse of the force in 1 minute,
we obtain

$$400v = \cdot 16 \times 30g \times 60 \text{ using 1 ton as unit of mass,}$$
therefore $v = 23 \cdot 04$ f.s. $= 15 \cdot 7$ miles per hour.

3-2

4·52. Problems on the running of trains with constant propulsive force, road resistance and brake resistance may conveniently be solved by equating the momentum to the impulse of the forces, and equating the kinetic energy to the work done, without introducing the question of acceleration.

Thus if m be the mass, P the propulsive force, and R_1 the road resistance, the equations for getting up a speed v in a distance x and time t from rest are

the impulse $(P - R_1)\, t = mv$ the momentum generated,

and the work done $(P - R_1)\, x = \frac{1}{2} mv^2$ the kinetic energy created.

Similarly, after steam is shut off and brakes are applied the total resistance to motion is $R_1 + R_2$ and if this brings the train to rest in a distance x' and time t', then since

momentum destroyed = impulse of the retarding force,

therefore $\qquad\qquad mv = (R_1 + R_2)\, t'$;

and since

kinetic energy destroyed = work done by the retarding force,

therefore $\qquad\qquad \frac{1}{2} mv^2 = (R_1 + R_2)\, x'$.

4·53. Example. *A train can be accelerated by a force of 55 lb. per ton weight and when steam is shut off can be braked by a force of 440 lb. per ton weight. Find the least time between stopping stations 3850 ft. apart, the greatest velocity of the train and the horse-power per ton weight necessary for the engine.*   [M. T. 1912]

In this case road resistance is neglected and with the notation of the last Article we take $m = 2240$, $P = 55g$, $R_2 = 440g$ and $x + x' = 3850$; and the equations are

$$55gt = 2240v = 440gt'$$

and $\qquad\qquad\quad 55gx = \frac{1}{2} \times 2240\, v^2 = 440gx'$,

therefore $\qquad\qquad x = \frac{7}{11} v^2$ and $x' = \frac{7}{88} v^2$.

Consequently $\qquad\quad \frac{63}{88} v^2 = 3850$, giving $v = 73\frac{1}{3}$ f.s.

Again $\qquad\qquad\qquad t = \frac{14}{11} v$ and $t' = \frac{7}{44} v$,

therefore $\qquad\qquad t + t' = \frac{63}{44} v = 105$ secs.

The maximum horse-power per ton weight is $Pv/550g = 7\frac{1}{3}$.

4·54. Effectiveness of Brakes. The effect of applying brakes to one pair of wheels of a four-wheeled car in motion is to alter the division of the weight between the wheels. Let the centre of gravity G be at a height h above the road and at horizontal distances a from the rear axle and a' from the front axle.

Neglecting the inertia of the wheels the friction forces

between them and the ground are negligible when they are running freely.

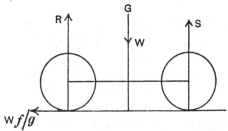

Let W be the weight of the car and suppose that a retardation f is caused by applying brakes to the rear wheels. This means that friction to an amount Wf/g must act on the back wheels opposing the motion of the car. Let R and S be the vertical reactions of the ground on the rear and front wheels.

By resolving vertically we have

$$R + S - W = 0.$$

Neglecting the rotatory inertia of the wheels and taking moments about G^* we get

$$aR - a'S + hWf/g = 0.$$

Hence $R = W \dfrac{a'g - hf}{g(a + a')}$ and $S = W \dfrac{ag + hf}{g(a + a')}$;

whereas if the brakes were not in use we should have

$$R : S = a' : a.$$

The maximum retardation that can be produced by applying brakes to the rear wheels without causing skidding is found from the condition

$$Wf/g \not> \mu R,$$

i.e. $$f(a + a') \not> \mu(a'g - hf)$$

or $$f \not> \mu a'g/(a + a' + \mu h).$$

If on the other hand brakes are applied to all four wheels the friction force Wf/g will be divided between front and back wheels, but the two equations for R and S will remain the same and the maximum retardation will be given by the condition

$$Wf/g \not> \mu(R + S),$$

or $$f \not> \mu g.$$

* The justification for this step will appear in Chapter ix.

4·55. Motion on an inclined plane. If a body of mass m is placed on an inclined plane which makes an angle α with the horizontal, its weight mg can be resolved into components $mg\cos\alpha$ at right angles to the plane and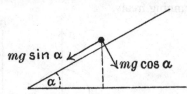

$mg\sin\alpha$ down the plane. The former component represents the pressure of the body on the plane and the latter component $mg\sin\alpha$ will, if there be no other force along the plane, cause an acceleration $g\sin\alpha$ down the plane. Consequently a particle freely projected up or down the plane has an acceleration $g\sin\alpha$ down the plane.

If however there is friction between the particle and the plane with coefficient μ, since the normal pressure is $mg\cos\alpha$, therefore there is a frictional force $\mu mg\cos\alpha$ opposing the motion. Hence if the particle moves up the plane it has a retardation $g(\sin\alpha+\mu\cos\alpha)$, but if it moves down the plane it has an acceleration $g(\sin\alpha-\mu\cos\alpha)$.

4·56. Example. *A train of* 100 *tons is ascending an incline of* 1 *in* 224 *with an acceleration of* 1 *f.s.s. Find the resistance to motion in pounds weight per ton of the train if at the speed of* 15 *miles per hour the horse-power developed is* 360.

Since the speed of 15 m.p.h. $=22$ f.s., therefore if P is the propulsive force in poundals,
$$22P=\text{rate of working}=360\times550g,$$
therefore $P=9000g$. The equation of motion is
$$mf=P-\tfrac{1}{224}mg-R,$$
where R is the total resistance in poundals.

Therefore
$$R=9000g-\tfrac{224000}{224}g-224000$$
$$=8000g-224000$$
$$=(8000-7000)\,g=1000\text{ lb. weight}$$
$$=10\text{ lb. weight per ton of the train.}$$

4·6. Resistance depending on velocity. When a body of mass m moves subject to a resistance proportional to some power of the velocity v, e.g. a resistance kv^2, we have an equation of motion
$$m\frac{dv}{dt}=P-kv^2,$$
where P is the force in the direction of motion.

When P is constant, this may be integrated by first writing it in the form

$$\frac{mdv}{P - kv^2} = dt,$$

which gives on integration

$$\frac{m}{2\sqrt{(Pk)}} \log \frac{\sqrt{P} + v\sqrt{k}}{\sqrt{P} - v\sqrt{k}} = t + \text{const.}$$

Alternatively, we may connect v with distance x, by writing

$$mv \frac{dv}{dx} = P - kv^2,$$

therefore

$$\frac{mdv^2}{P - kv^2} = 2dx,$$

and on integration $-\dfrac{m}{k} \log (P - kv^2) = 2x + \text{const.}$

If on the other hand the force P be not constant, but be working at a constant rate R, we have $Pv = R$ where R is constant.

Therefore

$$m \frac{dv}{dt} = \frac{R}{v} - kv^2,$$

or

$$mv \frac{dv}{dt} = R - kv^3.$$

This may be integrated by substituting $\dfrac{1}{z}$ for v, which leads to

$$t = \int \frac{mdz}{k - Rz^3} + \text{const.},$$

and is then integrable by partial fractions.

A simpler result is obtained connecting v and the distance x, for in this case

$$mv \frac{dv}{dx} = \frac{R}{v} - kv^2,$$

so that

$$\frac{mdv^3}{R - kv^3} = 3dx,$$

leading to

$$-\frac{m}{k} \log (R - kv^3) = 3x + \text{const.}$$

4·61. Example. *The resistance to the motion of a car varies as the square of its speed and the effective horse-power exerted at the road wheels is constant and equal to* 22. *If the weight of the car is* 1 *ton and the maximum speed of the car against its resistance is* 45 *miles an hour, determine the distance in which the car can accelerate from* 15 *to* 30 *miles per hour.* [M. T. 1917]

The equation of motion is

$$mv\,dv/dx = P - kv^2, \text{ where } m = 2240 \text{ lb.,}$$

and the rate of working $Pv = 22$ horse-power

$$= 22 \times 550g \text{ foot-poundals per second,}$$

therefore $\qquad P = 12100g/v = n/v$ say.

Hence $mv^2 dv/dx = n - kv^3$, so that by integrating we get

$$-\frac{m}{3k} \log (n - kv^3) = x + C, \text{ where } C \text{ is a constant} \quad \ldots\ldots\ldots(1).$$

Now at the maximum speed of 45 m.p.h. or 66 f.s. there is no acceleration, so that $v\,dv/dx = 0$ and $n = k \cdot 66^3$, but $n = 12100g$, therefore

$$k = 12100 \times 32/66^3 = 400/297.$$

Again the initial and final speeds are 15 m.p.h. and 30 m.p.h. or 22 f.s. and 44 f.s., and in (1) we may find C by putting $x = 0$ when $v = 22$, and then find the value of x when $v = 44$, namely

$$x = -\frac{m}{3k} \log \frac{n - k \cdot 44^3}{n - k \cdot 22^3}$$

$$= -\frac{m}{3k} \log \frac{66^3 - 44^3}{66^3 - 22^3} = -\frac{m}{3k} \log \frac{27 - 8}{27 - 1}$$

$$= 2240 \times \frac{99}{400} \log_e \frac{26}{19}$$

$$= 173 \cdot 9 \text{ ft. by the help of tables.}$$

4·62. Fall of a heavy body in a resisting medium. Suppose that the resistance is proportional to v^n, where v denotes the velocity. Measuring x vertically downwards, we may write

$$v \frac{dv}{dx} = g - kv^n.$$

The acceleration therefore decreases as the velocity increases, and the acceleration vanishes when the velocity attains the value $(g/k)^{\frac{1}{n}}$. From this instant the velocity will remain constant.

Since $(g/k)^{\frac{1}{n}}$ is the greatest velocity attained it is called *the limiting or terminal velocity.*

4·7. Motion of a Chain. The principles developed so far can be applied to the rectilinear motion of a uniform chain. Thus suppose a chain of length l and mass m per unit length to be placed on a smooth horizontal table in a line at right angles to the edge of the table, having at time t a length x hanging over the edge. Let v denote the velocity at time t, and let T denote the tension at the edge of the table. The equations of motion for the horizontal and vertical portions of the chain considered separately are

$$m\,(l - x)\,v\,dv/dx = T,$$

and
$$mxv\,dv/dx = mgx - T\,;$$

so that by adding we get

$$mlv\,dv/dx = mgx,$$

and on integration $lv^2 = gx^2 + C$, where C is a constant depending on the initial conditions.

This result may also be obtained by the principle of conservation of energy, for the kinetic energy of the whole chain is $\frac{1}{2}mlv^2$, and, if we take the zero level for the potential energy to be the upper surface of the table, the potential energy is $-mgx.\frac{1}{2}x$. Then, since the sum of the kinetic and potential energies is constant,

$$\tfrac{1}{2}mlv^2 - \tfrac{1}{2}mgx^2 = \text{const.},$$

which is equivalent to the last result.

4·71. Fall of a chain on to a table. A uniform fine chain of length l is suspended with its lower end just touching a horizontal table and allowed to fall; to find the pressure on the table when a length x has reached it.

Let m be the mass of unit length. When the velocity is v a length $v\delta t$ is brought to rest in a short interval of time δt; i.e. a mass $mv\delta t$ is brought to rest and therefore the amount of momentum destroyed is $mv^2\delta t$ in time δt. Hence the pressure on the table due to the rate of destruction of momentum is mv^2. But, since the upper end falls freely, the velocity when a length x has reached the table is given by $v^2 = 2gx$; and since the weight on the table is mgx, therefore the total pressure on the table is $3mgx$. This only holds good so long as x is less than l; as soon as x becomes equal to l the force due to impacts ceases and the pressure becomes mgl.

4·8. Units and Dimensions. The fundamental units are those of mass, space and time, denoted by **M**, **L** and **T**. All other dynamical units are derived from and expressible in terms of these in the same way as we expressed the units of velocity and acceleration in **3·4**; the dimensions of any physical quantity in terms of mass, space and time being indicated by the indices attached to **M**, **L** and **T**. Thus *momentum* being the product of mass and velocity, the unit of momentum is denoted by $\mathbf{MLT^{-1}}$. *Force* is rate of change of momentum or the product of mass and acceleration, so the unit of force is denoted by $\mathbf{MLT^{-2}}$.

Impulse is force multiplied by time, so that the unit of impulse is denoted by $\mathbf{MLT^{-1}}$, like the unit of momentum, since an impulse is equal to the change of momentum it produces. *Work or potential energy* is force multiplied by distance. Hence the unit of work is $\mathbf{ML^2\,T^{-2}}$; and the unit of *kinetic energy* has the same dimensions either because kinetic energy is equivalent to work done or because it is measured by $\frac{1}{2}\,mv^2$.

Power is rate of doing work or work divided by time. Therefore the unit of power is denoted by $\mathbf{ML^2\,T^{-3}}$.

4·81. Change of Units. If a certain physical quantity is of μ dimensions in mass, λ dimensions in length and τ dimensions in time, and V is its measure when **M, L, T** are the units of mass, length and time, then if the units of mass, length and time are changed to $\mathbf{M_1, L_1, T_1}$ the measure of the quantity becomes V_1 where

$$V_1 \mathbf{M_1}^\mu \mathbf{L_1}^\lambda \mathbf{T_1}^\tau = V \mathbf{M}^\mu \mathbf{L}^\lambda \mathbf{T}^\tau,$$

since these are equivalent representations of the same thing.

4·82. The consideration of dimensions is a useful check in dynamical work, for each side of an equation must represent the same physical thing and therefore must be of the same dimensions in mass, space and time.

Sometimes a consideration of dimensions alone is sufficient to determine the form of the answer to a problem. For example, suppose that a particle starts from rest at a distance a from a fixed origin O and moves with an acceleration μ/x^2 towards O, where x denotes distance from O. The time taken to arrive at O can only depend on the given constants a and μ. Suppose that the expressions for the time contain as factors a^α and μ^β. Then $a^\alpha\mu^\beta$ is of one dimension in time; but μ/x^2 is acceleration, i.e. of dimensions $\mathbf{L}\mathbf{T}^{-2}$ and x being of dimensions $\mathbf{L}$, it follows that μ is of dimensions $\mathbf{L}^3\mathbf{T}^{-2}$. Hence $a^\alpha\mu^\beta$ is of dimensions $\mathbf{L}^{\alpha+3\beta}\mathbf{T}^{-2\beta}$, but it is also of dimensions $\mathbf{T}$, therefore $\beta = -\frac{1}{2}$ and $\alpha = \frac{3}{2}$. Consequently the time taken to arrive at O is proportional to $a^{\frac{3}{2}}\mu^{-\frac{1}{2}}$.

EXAMPLES

[In numerical work take g to be 32 f.s.s.]

Uniform Acceleration

1. Find the minimum horse-power that a fire engine must have if it is to project 150 lb. of water per second with an initial velocity of 100 feet per second.   [S. 1911]

2. If a train travelling at 30 miles an hour picks up 10000 lb. of water in a quarter of a mile, find the back-pressure thereby produced on the train and the extra horse-power required on this account to maintain the speed.   [S. 1910]

3. A column of water 30 feet long is moving behind a plug piston in a pipe of uniform diameter with a velocity of 15 feet per second. Prove that the time average of the pressure of the water on the piston, caused by its stoppage in one-tenth of a second, is 61·0 lb. per square inch.
[S. 1915]

4. Shew that the horse-power required to pump 1000 gallons of water per minute from a depth of 50 feet, and deliver it through a pipe of cross-section 6 square inches, is about $34\frac{1}{2}$. (Assume that 1 cubic ft. of water is $6\frac{1}{4}$ gallons, and that 1 gallon of water weighs 10 lb. and neglect the friction losses.)   [S. 1912]

5. Find the horse-power required to lift 1000 gallons of water per minute from a canal 20 feet below and project it from a nozzle of cross-section 2 sq. inches.   [S. 1926]

6. An engine of 400 horse-power is drawing a train of 200 tons mass up an incline of 1 in 280 at 30 miles per hour; determine the road resistance in pounds weight per ton mass.   [S. 1910]

7. An engine working at 500 horse-power pulls a train of 200 tons along a level track, the resistances being 16 lb. per ton. When the velocity of the train is 30 miles per hour, find its acceleration.

At what steady speed will the engine pull the train up an incline of 1 in 100 with the same expenditure of power against the same resistances?

[S. 1917]

8. An engine of weight W tons can exert a maximum tractive effort of P tons weight and develop at most H horse-power. The resistances to motion are constant and equal to R tons weight. Shew that starting from rest the engine will first develop its full horse-power when its velocity is $\dfrac{55H}{224P}$ f.s. after at least $\dfrac{55\,WH}{224Pg\,(P-R)}$ seconds.

What is the greatest velocity which the engine can attain?   [S. 1923]

9. A car weighing 3 tons will just run down a slope of angle $\alpha\,(=\sin^{-1}\tfrac{1}{20})$ under its own weight. Assuming that the forces resisting its motion remain constant, and that the engine exerts a constant tractive force, find to the nearest unit the horse-power of its engine if it can attain a velocity of 30 miles per hour in 4 minutes on the level.   [S. 1922]

10. An engine driver of a train at rest observes a truck moving towards him down an incline of 1 in 60 at a distance of half a mile. He immediately starts his train away from the truck at a constant acceleration of 0·5 ft./sec.^{2}. If the truck just catches the train find its velocity when first observed. Assume that friction opposing the truck's motion is 14 lb. weight per ton.   [S. 1924]

11. The weight of a train is 200 tons, the part of the weight of the engine supported by the driving wheels is 25 tons and the coefficient of friction between the driving wheels and the rails is ·18. Prove that at the end of a minute after starting on the flat the velocity will be less than $29\tfrac{5}{11}$ miles per hour.   [S. 1911]

12. A particle is projected directly up a plane inclined at an angle α to the horizon, with initial velocity u given by

$$u^2 = 2gh\,(\sin\alpha + \mu\cos\alpha),$$

where μ is the coefficient of friction. Shew that it traverses a distance h, and that it stays at the highest position if $\tan\alpha < \mu$. If $\tan\alpha > \mu$, find its velocity when it returns to the starting point, and explain the difference between the initial and final kinetic energies.   [S. 1923]

13. A train of 200 tons, uniformly accelerated, acquires in two minutes from rest a velocity of 30 m.p.h. Shew that, if the coefficient of friction be ·18, the part of the load carried by the driving wheels of the engine cannot be less than 12·7 tons.   [S. 1921]

14. The tractive effort of an electric train is uniform and equal to the weight of 4 tons. The road resistance is 40 lb. weight per ton of the train, and the brake resistance is an additional 200 lb. weight per ton. The train is taken from one station to the next, distant half a mile, in $1\frac{1}{2}$ minutes, full power being kept on until the speed reaches 30 miles an hour, when the train 'coasts' at a uniform speed until power is shut off and the brakes are put on. Shew that the mass of the train is approximately 85 tons. [S. 1917]

15. An engine weighing 96 tons, of which 40 tons are carried by the driving wheels, exerting a uniform pull gives a train a velocity of 25 miles per hour after travelling for 50 seconds from rest against a resistance of 10·5 lb. weight per ton. If the friction between the driving wheels and the rails is 0·2 times the pressure, find the tension in the coupling between the engine and the first carriage. [S. 1923]

16. A train weighs 200 tons and the engine exerts a constant pull of 45 lb. per ton, resistance to motion being 10 lb. per ton. The train starts from rest; after a certain time steam is turned off and the brakes put on. The train comes to rest at a distance of 1050 yards from the starting point 2 mins. 20 secs. after it started. Find the retarding force per ton of the brakes, and also the greatest horse-power developed. [S. 1915]

17. A train travels from rest to rest between two stations 5 miles apart. The total mass is 200 tons; there is a road resistance of 12 lb. weight per ton, and the engine exerts a uniform pull of 5 tons weight until the maximum speed of 30 miles per hour is reached. This speed is maintained until, steam being shut off, an *additional* resistance equal to ·075 the weight of the train is applied to bring the train to rest. Find the time between the stations. [S. 1918]

18. A motor-car has its centre of gravity at a height h ft. midway between the axles, the wheel-base being l ft. Shew that the ratio between the least distances in which the car can be stopped by brakes acting (a) on the front wheels, (b) on the back wheels, is $\dfrac{l-\mu h}{l+\mu h}$, where μ is the coefficient of adhesion between the tyres and the ground. The rotary inertia of the wheels may be neglected. [S. 1917]

19. A mine cage, weighing with its load 5 cwt., is raised by an engine which exerts a constant turning moment on the rope drum which is 16 ft. in diameter. The speed rises until the engine is running at 60 revolutions per minute, when its output is 55 horse-power.

Find the acceleration and the time that elapses before the cage reaches full speed: also find how far the cage rises in that time. [S. 1915]

20. An engine moves at a steady velocity v along level ground when working at a constant horse-power H. When moving up a plane inclined at a small angle to the horizontal its steady velocity under the same horse-power is v'. If the engine starts down the same incline with velocity v' and moves for t seconds with a constant acceleration until it reaches its steady velocity down the plane corresponding to the same horse-power, H, shew that the distance travelled in these t seconds is $v'^2 t/(2v' - v)$. Assume that the frictional resistance is constant throughout.   [S. 1926]

21. A 20 horse-power motor lorry, weighing 5 tons, including load, moves up a hill with a slope of 1 in 20. The frictional resistance is equivalent to 13 lb. weight per ton, and may be supposed independent of the velocity. Find the maximum steady rate at which the lorry can move up the slope, and the acceleration capable of being developed when it is moving at 6 miles per hour.   [S. 1923]

22. A load W is to be raised by a rope, from rest to rest, through a height h; the greatest tension which the rope can safely bear is nW.

Shew that the least time in which the ascent can be done is $\left\{\dfrac{2nh}{(n-1)g}\right\}^{\frac{1}{2}}$.

[S. 1926]

Variable Acceleration

23. A tramcar starts from rest and its velocities at intervals of 5 secs. are given in the following table :

Time in seconds	0	5	10	15	20	25	30
Velocity in miles per hour	0	8·1	11·8	14·6	16·3	17·7	19

Calculate the distance in yards travelled in the above time. Also, if the car weigh 8 tons, estimate the effective pull exerted on the car at the end of 20 seconds.   [S. 1910]

24. A car whose mass is 2000 lb. starts from rest; the resistance to the motion is equal to 50 lb. weight. When it has travelled S feet the force exerted by the engine is P lb. weight where

S	0	10	20	30	40	50	60	70	80	90	100
P	644	634	622	607	587	565	537	509	475	440	404

Find, approximately, the velocity after the car has travelled 100 feet.

[M. T. 1914]

25. A body of mass 1 lb. is projected on a rough plane surface with a velocity of 10 feet per second, and its velocity after time t is given for various values of t by a smooth curve passing through the points defined by the following table :

Time in seconds	0	5	10	15	20	25	30
Velocity in feet per second	10·0	9·0	8·2	7·4	6·7	6·0	5·4

Derive the curve connecting the retardation with the distance travelled, and by determining the area of this curve verify that the energy lost during the 30 seconds is 1·1 ft.-lb. approximately. [S. 1915]

26. The speed of a motor cycle is observed, as it passes five posts placed 50 yards apart on a level track, to be 14·0, 26·4, 33·3, 37·1, 38·9 miles an hour respectively. Assuming the resistance in pounds weight due to mechanical and air friction to be $6 + 0·02v^2$, where v is expressed in miles an hour, calculate the horse-power actually developed by the engine when the speed is 35 miles an hour, the total weight of machine and rider being 400 lb. [S. 1921]

27. A locomotive of mass m tons starts from rest and moves against a constant resistance of P pounds weight. The driving force decreases uniformly from $2P$ pounds weight at such a rate that at the end of a seconds it is equal to P. Find the velocity and the rate of working after t seconds ($t < a$) and shew that the maximum rate of working is $1·54 \times 10^{-5} aP^2/m$ horse-power. [S. 1924]

28. The acceleration of a tramcar starting from rest decreases by an amount proportional to the increase of speed, from 1·5 f.s.s. at starting to 0·5 f.s.s. when the speed is 5 m.p.h. Find the time taken to reach 5 m.p.h. from rest. [S. 1925]

29. A train of weight M lb. moving at v feet per second on the level is pulled with a force of P lb. against a resistance of R lb. Shew that in accelerating from v_0 to v_1 feet per second, the distance in feet described by the train is $\dfrac{M}{g} \displaystyle\int_{v_0}^{v_1} \dfrac{v\,dv}{P-R}$. If the resistance $R = a + bv^2$, find an expression for the distance described when the power P is shut off and the velocity decreases from v_0 to v_1. [S. 1925]

30. Shew that a motor-car, for which the retarding force at V miles an hour when the brakes are acting may be expressed as $(1000 + 0·08 V^2)$ pounds weight per ton of car, can be stopped in approximately 57 yards from a speed of 50 miles an hour. [$\log_e 10 = 2·30$.] [S. 1927]

·31. The resistance of the air to bullets of given shape varies as the square of the velocity and the square of the diameter, and for a particular bullet (diameter 0·3″) is 40 times the weight at 2000 f.s. For an exactly similar bullet of the same material (diameter 0·5″) shew that the velocity will drop from 2000 f.s. to 1500 f.s. in about 500 yards, assuming the trajectory horizontal. [$\log_e 10 = 2·30$.]   [S. 1923]

32. A motor-bicycle which with its rider weighs 3 cwt. is found to run at 30 miles per hour up an incline of 1 in 20 and at 50 miles per hour down the same incline. Assuming that the resistance is proportional to the square of the velocity and that the engine is working at the same horse-power, find the speed that would be attained on the level, and shew that the horse-power is 2⅓ nearly.   [S. 1924]

33. The resistance to an aeroplane when landing is $a + bv^2$ per unit mass, v being the velocity, a, b constants. For a particular machine, $b = 10^{-3}$ ft.-lb.-sec. units and it is found that if the landing speed is 50 miles per hour the length of run of the machine before coming to rest is 150 yards. Calculate the value of the constant a.   [S. 1927]

34. A particle is projected vertically upwards with velocity V, and the resistance of the air produces a retardation kv^2, where v is the velocity. Shew that the velocity V' with which the particle will return to the point of projection is given by

$$\frac{1}{V'^2} = \frac{1}{V^2} + \frac{k}{g}.$$ [S. 1925]

35. OAB is a vertical circle of radius a. O is its highest point ; OA subtends angle α at the centre ; AB subtends angle 2β. ($\alpha + \beta < \frac{1}{2}\pi$.) Shew that the time taken for a particle to slide down the chord AB from rest at A is $2\sqrt{(a \cos \alpha/g)}$, when the angle of friction is also α.

Shew that if the motion is also subject to a resistance proportional to the velocity, the time of descent is still independent of β.   [S. 1926]

36. In starting a tram of mass 3200 lb. the pull exerted by a horse is initially 200 lb. and this pull decreases uniformly with the time until at the end of 10 secs. it has fallen to 40 lb., an amount just sufficient to overcome the frictional resistance of the tram. Shew by means of a curve the variation in the velocity and find the distance run during this period of 10 secs.

Shew that the horse-power is a maximum at the end of 5 secs., and find its maximum value.   [M. T. 1913]

37. In starting a train the pull of the engine on the rails is at first constant, and equal to P ; and after the speed attains a certain value u, the engine works at a constant rate $R = Pu$. Prove that when the engine

has attained a speed v greater than u, the time t and the distance x from the start are given by

$$t = \tfrac{1}{2}\frac{M}{R}(v^2 + u^2), \quad x = \tfrac{1}{3}\frac{M}{R}(v^3 + \tfrac{1}{2}u^3),$$

where M is the mass of the engine and train together.

Calculate the time occupied in attaining a speed of 45 miles an hour, when the total mass is 300 tons, the engine has 420 horse-power and can exert a pull equal to 12 tons weight.   [M. T. 1914]

38. The horse-power required to propel a steamer of 10,000 tons displacement at a steady speed of 20 knots is 15,000. If the resistance is proportional to the square of the speed, and the engines exert a constant propeller thrust at all speeds, find the acceleration when the speed is 15 knots.

Shew that the time taken from rest to acquire a speed of 15 knots is about $1\tfrac{1}{2}$ minutes, given $\log_e 7 = 1\cdot946$, one knot $= 100$ ft. per min.
                                                              [M. T. 1916]

39. The resistance to the motion of a train for speeds between 20 and 30 miles per hour, may be taken as $\frac{1}{400}V^2 + 9$ in pounds weight per ton, where V is the velocity in miles per hour. Sketch a curve shewing how the horse-power per ton, necessary to overcome the resistance, increases with the speed as the speed rises from 20 to 30 miles per hour, the train being on the level.

Steam is shut off when the speed is 30 miles per hour and the train slows down under the given resistance. In what time will the speed fall to 20 miles per hour?   [M. T. 1918]

40. A locomotive drawing a total weight of 264 tons on the level is exerting a tractive force of 20,000 pounds weight at the speed of 15 miles per hour. It works at constant horse-power until its speed is 60 miles per hour, when it is just able to overcome the resistance to motion, which may be taken to vary as the square of the velocity. Shew that it reaches the speed of 45 miles per hour from the speed of 15 miles per hour in a distance of approximately 5080 feet.   [M. T. 1919]

41. A train of mass 300 tons is originally at rest on a level track. It is acted on by a horizontal force F which uniformly increases with the time, in such a way that $F=0$ when $t=0$, $F=5$ when $t=15$; F being measured in tons weight, t in seconds. When in motion the train may be assumed to be acted on by a frictional force of 3 tons, independent of the speed of the train. Find the instant of starting, and shew that at $t=15$ the speed of the train is $0\cdot64$ foot per second, whilst the horse-power required at this instant is about 13.   [M. T. 1920]

42. A train of weight W lb. moving at V feet per second on the level is pulled with a force of F lb. against a train resistance of R lb. Shew that in accelerating from V_0 to V_1 ft. per sec., the distance in feet described by the train is

$$\frac{W}{g} \int_{V_0}^{V_1} \frac{V\,dV}{F-R}.$$

If $W = 300$ *tons*, $R = 2160 + 15\,V^2$, shew that the distance described in slowing down on the level from 45 to 30 miles per hour, with the power shut off, is about 537 feet. [$\log_e 10 = 2 \cdot 303$.]　　　　[M. T. 1921]

43. A cyclist works at the constant rate of P horse-power. When there is no wind, he can ride at 22 feet per second on level ground, and at 11 feet per second up a hill making an angle $\sin^{-1} \frac{1}{20}$ with the horizon. The total mass of man and cycle is 180 lb. The resistance of the air is kv^2 lb. weight, when the velocity of the man relative to the air is v feet per second ; the other frictional forces are negligible.

Find P, and shew that the speed of the cyclist when riding on level ground against a wind of 22 feet per second is between 10 and 10·5 feet per second.　　　　[M. T. 1924]

44. The external resistance to the motion of a bicycle and rider may be supposed to consist of two parts, a constant force and a force varying as the square of the speed. A rider observes that his speed when free-wheeling down a hill of slope 1 in 50 is sensibly constant when it reaches 10 feet per second, and that on a slope of 1 in 25 it becomes constant at 20 feet per second. The mass of the bicycle and rider is 200 lb. Find the power expended by the rider in maintaining a steady speed of 15 feet per second on the level, assuming that when the rider is propelling the bicycle 10 per cent. of the work he does is lost in internal friction in the pedalling gear.　　　　[M. T. 1926]

45. A uniform chain 30 centimetres long, having a mass of 1 gramme per centimetre, lies partly in a straight line along a rough horizontal table perpendicular to the edge. The portion hanging over the edge is just sufficient to cause the chain to begin to slip. The coefficient of friction with the table being $\frac{1}{2}$, find the velocity of the chain and its tension at the edge of the table when x centimetres have slipped off.　　　[M. T. 1919]

ANSWERS. $(g=32.)$

1. $42\frac{27}{44}$.   2. $458\frac{1}{3}$ lb. wt.; $36\frac{2}{3}$ H.P.   5. $180\frac{20}{33}$.   6. 17 lb. wt. per ton.   7. $2\frac{61}{80}$ f.s.s., $24\frac{53}{128}$ m.p.h.   8. $55H/224R$ f.s.   9. 30.

10. $\sqrt{880}$ f.s.   11. 2·25 f.s.s.   12. $\{2gh\,(\sin\alpha-\mu\cos\alpha)\}^{\frac{1}{2}}$.

15. 11984 lb. wt.   16. 53 lb. wt. per ton; $736\frac{4}{11}$ H.P.   17. 10 mins. $43\frac{5}{9}$ secs.   19. 2·39 f.s.s.; 21 secs.; 531 ft.   21. 12 m.p.h.; $1\frac{11}{14}$ f.s.s.   23. 191 yds.; 2546 lb. wt.   24. 40 f.s.   26. 3·4.

27. $P\,(t-t^2/2a)/70m$ f.s.; $P^2t\,(2-t/a)^2/140m$ ft. lb. per sec.

28. $\frac{32}{3}\log_e 3 = 8\cdot06$ secs.   29. $\dfrac{M}{2bg}\log\dfrac{a+bv_0^2}{a+bv_1^2}$.   32. $\frac{32}{3}\sqrt[3]{510}=58\cdot6$ f.s.

33. $\log_e\left(1+\dfrac{242}{45a}\right)=\cdot9$; $a=3\cdot68$.   36. $53\frac{1}{3}$ ft.; $1\frac{17}{56}$.   37. 3 mins. 21·35 secs.   38. $\frac{99}{640}$ f.s.s.   39. 97·13 secs.   41. $t=9$.   43. $P=\cdot2\ldots$

44. 93·5 ft. lb. per sec.   45. $7x$ cm. per sec.; $49\,(20-x)\,(10+x)$ gms. wt.

Chapter V

KINEMATICS IN TWO DIMENSIONS

51. We now proceed to consider the motion in one plane of a particle and of a rigid body, confining our considerations in this chapter to Kinematics or the geometry of the motion apart from the forces that cause it.

We require general definitions of velocity and acceleration applicable to curved paths.

The **velocity** of a point is defined to be a vector drawn through the point and such that its resolved part in any direction is the rate of displacement of the point in that direction.

Let the point be moving along a curve APQ, and let P, Q be its positions at times $t, t + \delta t$, and let $(x, y), (x + \delta x, y + \delta y)$ be the

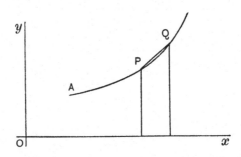

coordinate of P, Q referred to rectangular axes. Then the resolved parts of the velocity parallel to the axes are

$$\lim_{\delta t \to 0} \frac{\delta x}{\delta t} \text{ or } \frac{dx}{dt} \text{ or } \dot{x},$$

and

$$\lim_{\delta t \to 0} \frac{\delta y}{\delta t} \text{ or } \frac{dy}{dt} \text{ or } \dot{y}.$$

Again $\delta x, \delta y$ are the resolved parts of the vector PQ, so that the resultant velocity as defined above is

$$\lim_{\delta t \to 0} \frac{\operatorname{chd} PQ}{\delta t}.$$

But if s denotes the arc AP measured from a fixed point A on the curve and $s + \delta s$ denotes the arc AQ, we know that

$$\lim_{Q \to P} \frac{\operatorname{chd} PQ}{\delta s} = 1,$$

so that the resultant velocity is

$$\lim_{\delta t \to 0} \frac{\operatorname{chd} PQ}{\delta s} \cdot \frac{\delta s}{\delta t} = \frac{ds}{dt} \text{ or } \dot{s},$$

and its direction is along the tangent at P.

Further, if V denotes the resultant velocity and ψ its inclination to the axis of x, we have, as a verification, that the velocity parallel to

$$Ox = V \cos \psi = \frac{ds}{dt}\frac{dx}{ds} = \frac{dx}{dt} \text{ or } \dot{x},$$

and the velocity parallel to

$$Oy = V \sin \psi = \frac{ds}{dt}\frac{dy}{ds} = \frac{dy}{dt} \text{ or } \dot{y}.$$

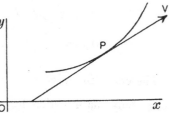

5·11. The **acceleration** of a point is a vector drawn through the point and such that its resolved part in any direction is the rate of change of velocity in that direction. (See **4·161.**)

Thus if we use rectangular axes and u, v denote the resolved parts of the velocity parallel to the axes at time t, and $u + \delta u$, $v + \delta v$ denote the resolved parts at time $t + \delta t$, then the resolved parts of the acceleration are

$$\lim_{\delta t \to 0} \frac{\delta u}{\delta t} = \frac{du}{dt} = \dot{u} = \ddot{x},$$

and

$$\lim_{\delta t \to 0} \frac{\delta v}{\delta t} = \frac{dv}{dt} = \dot{v} = \ddot{y}.$$

It does not follow however that the resultant acceleration is directed along the tangent to the path of the point.

Let $V, V + \delta V$ denote the velocities at the points P, Q of the path, where PQ is the small arc described in time δt, and let $\delta \psi$ denote the angle between the tangents at P, Q.

From any point O draw vectors $\overline{Oa}, \overline{Ob}$ to represent V and $V + \delta V$. Then $\overline{ab}$ represents the change of velocity in time δt.

Draw am at right angles to Oa meeting Ob in m. Then
$$\overline{ab} = \overline{mb} + \overline{am}.$$

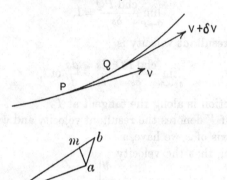

Therefore the acceleration, being $\lim\limits_{\delta t \to 0} \dfrac{\overline{ab}}{\delta t}$ both as regards magnitude and direction, has components

$\lim\limits_{\delta t \to 0} \dfrac{\overline{mb}}{\delta t} = \dfrac{dV}{dt}$ or $\dot{V}$ in the direction of the tangent to the path

and $\qquad \lim\limits_{\delta t \to 0} \dfrac{\overline{am}}{\delta t} = \lim\limits_{\delta t \to 0} \dfrac{V\delta\psi}{\delta t} = \lim\limits_{\delta t \to 0} \dfrac{V\delta\psi}{\delta s} \cdot \dfrac{\delta s}{\delta t} = \dfrac{V^2}{\rho}$,

where ρ is the radius of curvature of the path, and the direction of this component is that of am, i.e. along the inward normal to the path.

We note that the tangential component of acceleration dV/dt may be written $\dfrac{dV}{ds}\dfrac{ds}{dt}$ or $V\dfrac{dV}{ds}$.

5·12. Hence we have equivalent representations of velocity as shewn in the diagrams.

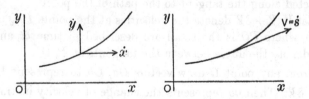

Also equivalent representations of acceleration as shewn in the diagrams

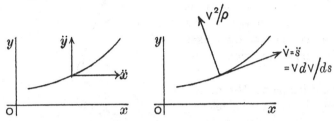

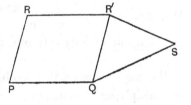

and we note that we do not represent acceleration by a single symbol unless the path is a straight one; in fact, bending of the path involves a normal component of acceleration.

5·2. Relative Velocity. Let two points P, Q move in the same plane with velocities represented by the vectors $\overline{PR}$, $\overline{QS}$. Complete the parallelogram $PQR'R$. Then $\overline{QR'}$ is a vector representing a velocity equal in magnitude and direction to that of P. But $\overline{QS} = \overline{QR'} + \overline{R'S}$, therefore $\overline{R'S}$ represents a velocity which when compounded with the velocity of P gives the velocity of Q. Hence $\overline{R'S}$ is the velocity of Q relative to P. Similarly $\overline{SR'}$ is the velocity of P relative to Q.

We may also represent the same thing analytically. Thus if (x, y), (x', y') are the coordinates of P, Q referred to any frame of axes in a plane, we may call
$$\xi = x' - x \quad \text{and} \quad \eta = y' - y$$
the coordinates of Q relative to P, and by differentiation we find the relative velocity components
$$\dot{\xi} = \dot{x}' - x' \quad \text{and} \quad \dot{\eta} = \dot{y}' - y'.$$

In other words the velocity of Q relative to P is to be found by subtracting the velocity of P from that of Q, either geometrically in the form of vectors, or algebraically by subtracting corresponding components.

5·3. Angular Velocity. If a point P is moving in a plane the angular velocity of P about a fixed point O in the plane is the rate of increase of the angle that the line OP makes with a fixed direction in the plane.

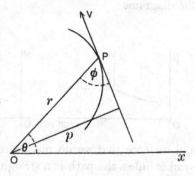

Thus if a point P be moving in any direction with velocity V, and OP makes an angle θ with a fixed direction Ox, the angular velocity of P about $O = d\theta/dt$. But if $OP = r$ and the tangent to the path makes an angle ϕ with OP we have

$$r\,d\theta/ds = \sin\phi,$$

therefore
$$\frac{d\theta}{dt} = \frac{ds}{dt}\frac{\sin\phi}{r} = \frac{V\sin\phi}{r}$$

or   angular velocity = (component velocity at

right angles to OP)/OP.

We may also write this result $r\dot\theta = V\sin\phi$,
or multiplying by r we get $r^2\dot\theta = Vp$,
where p is the perpendicular distance of the tangent from the origin. That is $r^2\dot\theta$ = moment of the velocity about the origin.

Again, a sectorial element of area of a plane curve is represented by $\frac{1}{2}r^2\delta\theta$, so that $\frac{1}{2}r^2\dot\theta$ is the rate of increase of a sectorial area as P moves along the curve, and therefore Vp is twice the rate at which the radius vector OP sweeps out area.

5·31. Examples. (i) *A point is moving in a circle. Shew that at any instant its angular velocity about a point on the circumference is half its angular velocity about the centre.*

This follows from the fact that in the diagram the angle
$$POC = \tfrac{1}{2}PCx$$
so that $\dfrac{d}{dt}(POC) = \tfrac{1}{2}\dfrac{d}{dt}(PCx).$

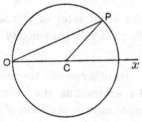

(ii) *A point moves in a parabola with constant velocity. Shew that the angular velocity about the focus varies as $\cos^3 \tfrac{1}{2}\theta$, where θ is the angular distance from the vertex.*

We have $\dot\theta = \dfrac{Vp}{r^2}$, but in the parabola with the focus as origin

$$p^2 = ar, \text{ so that } \dot\theta \propto \frac{1}{r^{\frac{3}{2}}};$$

but
$$\frac{l}{r} = 1 + \cos\theta = 2\cos^2 \tfrac{1}{2}\theta,$$

therefore
$$\dot\theta \propto \cos^3 \tfrac{1}{2}\theta.$$

5·32. Motion in a Circle. At time t let V be the velocity of
a point P moving in a circle of
radius r, centre O. Let θ be the
angle that OP makes with a
fixed direction Ox.

Let s be the arc AP measured
from a point A on Ox. Then
the velocity

$$V = \dot s = r\dot\theta.$$

Also the acceleration com-
ponents are $\ddot s$ or $r\ddot\theta$ along the
tangent and V^2/r or $r\dot\theta^2$ towards the centre.

If the motion be uniform, i.e. V constant, then there is of
course no acceleration along the tangent, but there is in every
case the acceleration V^2/r or $r\dot\theta^2$ towards the centre.

5·33. Relative Angular Velocity. If two points P, Q are
moving in a plane with velocities u, v, then either point in general
possesses an angular velocity relative to the other, which may be
represented thus :

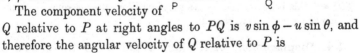

Let the directions of the
velocities u, v make angles
θ, ϕ with the line PQ.

The component velocity of
Q relative to P at right angles to PQ is $v\sin\phi - u\sin\theta$, and
therefore the angular velocity of Q relative to P is

$$(v\sin\phi - u\sin\theta)/PQ.$$

Whenever $v\sin\phi = u\sin\theta$, then the line PQ is moving parallel
to itself.

5·34. Example. *Two points describe concentric circles with velocities varying inversely as the square roots of the radii of the circles; to find positions in which the relative angular velocity vanishes.*

Let O be the centre of the circles, a, b the radii, and u, v the velocities. If P, Q are positions of the points when the relative angular velocity vanishes, the velocities resolved at right angles to PQ must be equal. Let $POQ = \theta$.

Therefore $u \cos OPQ + v \cos OQP = 0$,

or $a^{-\frac{1}{2}} \cos OPQ - b^{-\frac{1}{2}} \cos (\theta + OPQ) = 0$,

or $(a^{-\frac{1}{2}} - b^{-\frac{1}{2}} \cos \theta) \cos OPQ$
$\qquad\qquad + b^{-\frac{1}{2}} \sin \theta \sin OPQ = 0.$

But $\quad a \sin OPQ - b \sin OQP = 0$,

therefore $\qquad\qquad a \sin OPQ - b \sin (\theta + OPQ) = 0$,

or $\qquad\qquad (a - b \cos \theta) \sin OPQ - b \sin \theta \cos OPQ = 0.$

Eliminating the angle OPQ we get

$$(a^{-\frac{1}{2}} - b^{-\frac{1}{2}} \cos \theta)(a - b \cos \theta) + b^{\frac{1}{2}} \sin^2 \theta = 0,$$

therefore $\qquad\qquad \cos \theta = a^{\frac{1}{2}} b^{\frac{1}{2}} / (a - a^{\frac{1}{2}} b^{\frac{1}{2}} + b).$

This problem has a practical application. If we regard the points P, Q as representing the earth and an inferior planet describing circular orbits round the sun O in the same plane, we have found the relative positions of P, Q in their orbits at which the planet Q would appear to be 'stationary' as seen from the earth; since in this position there is no relative angular velocity. Since $v > u$, therefore when the planet Q is on the line OP, the join PQ must be turning in a clockwise sense; the motion of the planet among the stars as seen from the earth is then described as 'retrograde.' The position found above is that in which the clockwise rotation of PQ ceases and a counter-clockwise rotation is just about to begin. The visible motion of the planet then becomes 'direct,' and continues to be direct until the relative position is such a position as $Q'P$ in the diagram, where $Q'OP = \theta$, this being another 'stationary' position, after passing which the motion again becomes retrograde.

5·4. Displacement of a Plane Rigid Body in its Plane.

By a rigid body we understand a body in which the distance between any two particles remains invariable. The position in its plane of a plane rigid body is therefore determined when the positions of any two points A, B of the body are known. For if one point A of the body is fixed the only possible motion is a rotation about A, and if a second point B is fixed no motion is possible.

We shall now shew that *any displacement of such a body in its plane consists in a rotation about a point in the plane.*

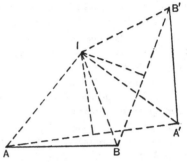

Let A, B be any two points of the body which become displaced to A', B'. Draw the perpendicular bisectors of AA', BB' meeting in I.

Then $IA = IA'$, $IB = IB'$ and $AB = A'B'$. Therefore the triangles $IAB, IA'B'$ are identically equal and the second is obtained from the first by a rotation about I through an angle AIA' or BIB'. I may be called *the centre of rotation* for this displacement.

If the displacement be a pure translation AA' and BB' are parallel and the point I is at an infinite distance.

5·41. Instantaneous Centre of Rotation. Let a plane body be moving in any manner in its plane, and consider the displacement that takes place in a short interval of time δt. Any two points A, B of the body have definite paths of motion and the displacements AA', BB' that take place in time δt are

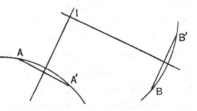

chords of small elements of these paths, and the point I, obtained as the intersection of the perpendicular bisectors of AA', BB', will, as $\delta t \to 0$, tend to coincidence with the intersection of the normals to the paths at A and B. Hence it follows that at any instant the body is turning about an *instantaneous centre of rotation I*, which is found by taking any two points A, B of the body and drawing normals to the paths of A and B through their instantaneous positions.

It follows that if P be the position of any other point of the body at the same instant, the direction of motion of P must be at right angles to IP; and, if $\delta\theta$ be the angle turned through by the body in time δt, the displacement of P in time δt is $IP.\delta\theta$.

5·42. Examples. (i) *The ends of a rod AB are constrained to move along two given lines Ox, Oy in the same plane.*

The instantaneous centre of rotation is the point I found by drawing AI, BI perpendicular to Ox, Oy respectively.

(ii) *A circular lamina rolls along a straight line. Prove that the instantaneous centre of rotation is the point of contact.*

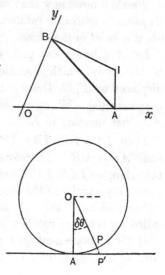

Let O be the centre, and A the point of contact. If by rolling through a small angle $\delta\theta$ the point P on the circle is brought into coincidence with P' on the straight line, then PP' is of order $(\delta\theta)^2$. Now since the path of O is parallel to the given line, therefore the instantaneous centre is on OA, and it is also on the limiting position as $\delta\theta \to 0$ of the perpendicular bisector of PP'. But $AP' = $ arc AP, and therefore as $\delta\theta \to 0$ the ratio $AP'/$chd $AP \to 1$, therefore this perpendicular bisector ultimately passes through A which is therefore the instantaneous centre of rotation. Another proof of this theorem will be given in **5·51**.

5·43. Pole Curves. When a lamina is in motion in its plane there is a locus of the instantaneous centre of rotation in the lamina and also a locus in the fixed plane. These loci are called *the pole curves*. Thus in the last example, of a circular lamina rolling along a straight line, the locus of the instantaneous centre in the lamina is the circular boundary, and the locus in space is the fixed straight line. And this example illustrates the general theorem that the motion of the lamina can be produced by the rolling of one pole curve on the other.

Without attempting a rigorous demonstration of this theorem, we can give an explanation that may satisfy the reader as to the truth of the theorem.

Let P, Q, R, S, T be a succession of points on the locus of the instantaneous centre in space. Consider the results of making infinitesimal displacements of the body about the points Q, R,

S, T in succession. The first of these displaces QP to QP_1, the second displaces RQP_1 to RQ_1P_2, the third displaces SRQ_1P_2 to

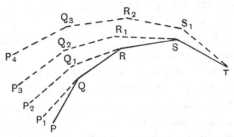

$SR_1Q_2P_3$, and the fourth displaces $TSR_1Q_2P_3$ to $TS_1R_2Q_3P_4$. Now S_1, R_2, Q_3, P_4 are a series of positions on the lamina of the instantaneous centre of rotation, i.e. they are points on the locus on the lamina of the instantaneous centre and they can in turn be brought back into coincidence with the point S, R, Q, P by the rolling of one polygon on the other. Hence by regarding the pole curves as the limits of polygons the motions of the body can be produced by rolling the pole curve fixed on the lamina on the pole curve fixed in space.

5·44. Example. *A triangular lamina ABC moves in its plane so that the sides AB, AC always pass through two fixed points P, Q. Find the pole curves.*

The point P on the lamina can only move along AB, for if it had any component of displacement at right angles to AB the side AB would no longer pass through the point P fixed in the plane.

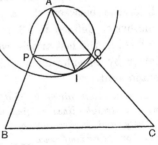

Hence the instantaneous centre I is on the perpendicular to AB through P, and in like manner on the perpendicular to AC through Q.

Therefore the angle PIQ is the supplement of A and a constant angle; and P, Q are fixed points, therefore the pole curve fixed in the plane is a circle PIQ.

Again AI is a diameter of this circle and is a fixed length (in fact $PQ \operatorname{cosec} A$). Hence the pole curve on the lamina is a circle of centre A and radius $PQ \operatorname{cosec} A$, i.e. twice the size of the former circle; and the motion of the triangle could be produced by rolling the larger circle on the smaller.

5·5. Angular Velocity of a Body. When a plane body is moving in its plane we may speak of the 'angular velocity of the body' without specifying any particular line or point of reference, meaning thereby the rate of increase of the angle between *any* line AB fixed in the body and *any* line Ox fixed in the plane of motion : i.e. if at time t the angle between AB and Ox be θ, then $\dot{\theta}$ is the angular velocity of the body. Also if any other line CD fixed in the body makes an angle $\theta + \alpha$ with Ox at the same instant, then α is the constant angle between AB and CD and unaltered by the motion, so that $\dot{\theta}$ is also the rate of increase of the angle between CD and Ox. Consequently in measuring the angular velocity of the body, it is immaterial upon what line fixed in the body we concentrate our attention.

Further, the angular velocity of the body is independent of any motion of translation of the body as a whole, for such a motion would not alter the direction of any line fixed in the body. For example, when we say that a wheel has an angular velocity ω in its plane this statement is independent of whether the wheel is turning about a fixed point, such as its centre, or rolling along in contact with a fixed line, or possessing any other translational motion, ω being the rate of increase of the angle between any line fixed in the plane of the wheel and a line fixed in the plane of motion.

5·51. It is important that the reader should appreciate the difference between the angular velocity of one point about another as defined in **5·3** and the angular velocity of a body as defined in the last article, we therefore propose to illustrate the latter by considering a simple problem from more than one standpoint.

A circle rolls along in contact with a straight line; to find the velocity and acceleration of any point on the circumference.

Take the given line as axis of x. Let C be the centre of the circle, a the radius, P the point of contact with Ox at time t, A the point on the circumference whose velocity and acceleration are re-

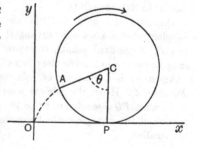

quired. Let θ be the angle ACP, then $\dot{\theta}$ is the angular velocity of the
circle. For convenience take the origin O so that $OP=\text{arc } AP$, i.e. O is
a position of the point A in the rolling motion. Taking an axis Oy at
right angles to Ox, let x, y be the coordinates of A, then

$$x = OP - AC\sin\theta = a(\theta - \sin\theta),$$
$$y = CP - AC\cos\theta = a(1 - \cos\theta).$$

Hence the components of velocity are

$$\dot{x} = a\dot{\theta}(1 - \cos\theta),$$

and
$$\dot{y} = a\dot{\theta}\sin\theta\,;$$

and the components of acceleration are

$$\ddot{x} = a\ddot{\theta}(1 - \cos\theta) + a\dot{\theta}^2\sin\theta,$$

and
$$\ddot{y} = a\ddot{\theta}\sin\theta + a\dot{\theta}^2\cos\theta.$$

As a special case we may put $\theta=0$, so that A is the point of contact of
the circle and the line; then we find that

$$\dot{x} = 0 \text{ and } \dot{y} = 0,$$

shewing that the point of contact of the rolling circle has no velocity, i.e.
it is the instantaneous centre of rotation as proved previously in **5·42**. In
the same case the acceleration components are $\ddot{x}=0$, $\ddot{y}=a\dot{\theta}^2$, shewing that
the point of contact of the rolling circle has an acceleration $a\dot{\theta}^2$ towards
the centre, as proved previously in **5·32.**

5·52. We may also obtain the foregoing expressions for velocity and
acceleration by compounding the velocity and acceleration relative to the
centre of the circle with the velocity
and acceleration of the centre. Thus,
let the circle be rolling with angular
velocity ω and let V be the velocity of
its centre C. Every point on the circumference has a velocity $a\omega$ relative
to the centre, so that the total velocity
of the point P on the circle is $V - a\omega$.
But assuming 'rolling' to mean that
P is the instantaneous centre of ro-
tation it follows that P has no velocity and therefore $V = a\omega$.

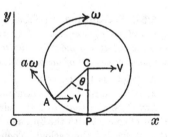

The velocity of a point A on the circumference is therefore compounded
of $a\omega$ along the tangent relative to the centre and V or $a\omega$ parallel to Ox,
giving as before components

$$a\omega(1 - \cos\theta) \text{ parallel to } Ox,$$

and
$$a\omega\sin\theta \text{ parallel to } Oy,$$

where $\omega = \dot{\theta}$.

Again for accelerations we have that the accelerations of A relative to
C are

$$a\dot{\omega} \text{ along the tangent and } a\omega^2 \text{ along } AC \text{ (\textbf{5·32}),}$$

and the acceleration of C is $\dot{V}$ parallel to Ox.

But $V = a\omega$, therefore $\dot{V} = a\dot{\omega}$.

Hence the accelerations of the point A are compounded of $a\dot\omega$ along the tangent, $a\omega^2$ along AC and $a\dot\omega$ parallel to Ox, and these are equivalent to components

$$a\dot\omega(1-\cos\theta)+a\omega^2\sin\theta \text{ parallel to } Ox,$$

and $\qquad a\dot\omega\sin\theta+a\omega^2\cos\theta$ parallel to Oy, as before.

5·53. We may also obtain the velocity components of the point A by considering the motion relative to the instantaneous centre of rotation P.

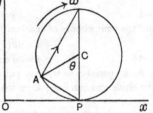

The point A is moving at right angles to AP with velocity ωAP, and since $AP=2a\sin\tfrac12\theta$, this gives components

$$2a\omega\sin^2\tfrac12\theta$$

or $\qquad a\omega(1-\cos\theta)$ parallel to Ox,

and $\qquad 2a\omega\sin\tfrac12\theta\cos\tfrac12\theta$

or $\qquad a\omega\sin\theta$ parallel to Oy,

as before.

But the finding of the accelerations of the point A by reference to the instantaneous centre would be more cumbersome than the process of **5·52**, because it would involve compounding accelerations of A relative to P with accelerations of P, and the latter accelerations would have to be found by compounding accelerations of P relative to C with the acceleration of C, as in **5·52**.

5·54. Examples. (i) *Prove, by considering a point on the circumference of a circle rolling uniformly along a straight line, that the radius of curvature of a cycloid at any point is twice the length of the line joining that point to the point of contact of the generating circle with the base.* [M. T. 1908]

A cycloid is the curve traced out by any point on the circumference of a circle which rolls along a straight line.

Let C be the centre of the circle, G the point of contact with the line, and P the point that is tracing the cycloid. Then since G is the instantaneous centre of rotation the point P is moving at right angles to PG with velocity $v=\omega PG$, where ω is the angular velocity of the circle. Now if ρ is the radius of curvature

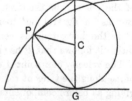

of the cycloid the normal component of the acceleration of P is v^2/ρ. But the acceleration of P may also be represented by its acceleration relative to C, i.e. $\omega^2 PC$ (**5·32**) compounded with the acceleration of C. But C has no acceleration because the motion is uniform. Therefore the resultant acceleration of P is $\omega^2 PC$ along PC.

Hence $\qquad v^2/\rho=\omega^2 PC\cos CPG=\tfrac12\omega^2 PG.$

But $\qquad v=\omega PG$, therefore $\rho=2PG.$

(ii) *A circle A of radius a turns round its centre with uniform angular velocity ω. A circle B of radius b rolls on the circle A and its uniform angular velocity is ω'. Find the time taken*

(1) *for the point of contact to make a complete circuit of A,*

(2) *for the centre of B to return to a former position.*

Determine the accelerations of the common point of the two circles and the greatest acceleration of a point on the circle B.

Suppose that at time $t=0$ the points M, N on the circles are in contact and on the fixed line Ax through the centre A of the circle A. Let P be the point of contact at time t. Since AM, BN are lines fixed in relation to the circles, in time t they turn through angles ωt, $\omega't$. Therefore if BN meets Ax in K, the angle $BKx=\omega't$ while $MAx=\omega t$. Again the arcs PN, PM are equal, so that if $PAM=\theta$ then the angle $PBN=a\theta/b$.

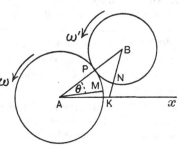

Therefore $\omega't=\omega t+\theta+a\theta/b$,

or $\theta=\dfrac{b(\omega'-\omega)t}{a+b}$ and $\dot\theta=\dfrac{b(\omega'-\omega)}{a+b}$.

And the time taken for P to make a circuit of the circle A is

$$2\pi/\dot\theta=2\pi(a+b)/b(\omega'-\omega).$$

Again the time taken for the point B to describe a circle round A depends on the angular velocity of the line AB, which $=\dot\theta+\omega$.

Hence the time required $=2\pi/(\dot\theta+\omega)$

$=2\pi(a+b)/(a\omega+b\omega')$.

For the accelerations we have that the acceleration of the point P on the circle A

$$=a\left(\frac{d}{dt}PAx\right)^2 \text{ along } PA \text{ by } \mathbf{5·32}$$

$$=a(\dot\theta+\omega)^2=a(a\omega+b\omega')^2/(a+b)^2\,;$$

and the acceleration of the point P on the circle B

$=$ accel. relative to centre $B+$ accel. of B

$=-b\omega'^2+(a+b)(\dot\theta+\omega)^2$ along BP

$=-b\omega'^2+(a\omega+b\omega')^2/(a+b)$.

Also the acceleration of any point on the circle B is compounded of its acceleration relative to the centre B, i.e. $b\omega'^2$ along the radius towards B, and the acceleration of B, so that the greatest acceleration will be when these components coincide in direction, i.e.

$$b\omega'^2+(a\omega+b\omega')^2/(a+b).$$

EXAMPLES

1. A ship sailing N.W. by compass through a tide running 5 knots finds that after 2 hours it has made 4 nautical miles S.W. Determine the direction of the current and the speed of the ship.

2. An aeroplane which travels at the rate of 80 miles per hour in still air starts from A to go to B which is 200 miles distant N.E. of A. If there is a wind blowing from the North at 20 miles per hour, determine the direction in which the aeroplane must move, and the time required.

If at the end of an hour the wind drops to 5 miles per hour, determine the position relatively to B of the aeroplane at the time when it should have arrived at B.

Prove that provided the velocity of the wind remains fixed in direction and magnitude, all points attainable by an aeroplane in a given time lie on a circle whose radius is independent of the wind.   [S. 1924]

3. The velocity of a stream between parallel banks at distance $2a$ apart is zero at the edges and increases uniformly to the middle where it is u. A boat is rowed with constant velocity $v\,(>u)$ relative to the water, and goes in a line straight across. How are the bows pointed at any point of the path and how long will it take to get across?   [S. 1921]

4. An aeroplane has a speed of v miles per hour, and a range of action (out and home) of R miles in calm weather. Prove that in a north wind of w miles per hour its range of action is

$$R\,(v^2 - w^2)/v\,(v^2 - w^2 \sin^2 \phi)^{\frac{1}{2}}$$

in a direction whose true bearing is ϕ. If $R = 200$ miles, $v = 80$ miles per hour and $w = 30$ miles per hour, find the direction in which its range is a maximum, and the value of the maximum range.   [M. T. 1921]

5. If a point moves so that its angular velocity about two fixed points is the same prove that it describes a circle.   [S. 1903]

6. If two particles describe the same circle of radius a, in the same direction with the same speed u, shew that at any instant their relative angular velocity is u/a.   [S. 1910]

7. A particle P is moving in a circle of radius a centre C with uniform speed u. AB is a diameter of the circle and $AP = r$. Find the angular velocity of P about A, B and C. What is the angular acceleration of P about the same points?   [S. 1909]

8. A particle P moves in an ellipse whose foci are S and H, and centre C. The velocity at any point of the path varies as the square of the diameter conjugate to CP. Prove that the angular velocity of P about S varies inversely as its angular velocity about H. [S. 1911]

9. The line joining two points A, B is of constant length a and the velocities of A, B are in directions which make angles α and β respectively with AB. Prove that the angular velocity of AB is $\dfrac{u \sin (\alpha - \beta)}{a \cos \beta}$, where u is the velocity of A. [S. 1918]

10. Two points are describing concentric circles of radii a and a' with angular velocities ω and ω' respectively. Prove that the angular velocity of the line joining them when its length is r is

$$\{(r^2 + a^2 - a'^2)\,\omega + (r^2 - a^2 + a'^2)\,\omega'\}/2r^2.$$ [S. 1913]

11. The end P of a straight rod PQ describes with uniform angular velocity a circle whose centre is O, while the other end Q moves on a fixed line through O in the plane of the circle. The end Q' of an equal straight rod PQ' moves on the same fixed line through O. Prove that the velocities of Q and Q' are in the ratio $QO : OQ'$. [S. 1925]

12. A circular ring of radius b turns round a fixed point O in its circumference with uniform angular velocity Ω. A smaller ring of radius a rolls on the inside of the larger ring with uniform angular velocity ω, the angular velocities being in the same sense. Find the velocity of any point of the smaller ring in any position. Also shew that, if $a\omega = b\Omega$, then in every position of the smaller ring one point on it is at rest. Indicate the position of this point for a general position of the rings. [S. 1921]

13. C is the centre of two concentric circles A, B, and a line CPQ meets the circles in P, Q. Tangents XPX, YQY are drawn to the circles at P, Q. The circle A rolls along the line XX carrying the circle B with it, so that C, P, Q are always collinear, until the point P is again on the line XX and Q is consequently again on the line YY. The distance between the two positions of P is equal to the circumference of the circle A. Investigate the fallacy in the assertion that the distance between the two corresponding positions of Q is equal to the circumference of the circle B. [S. 1924]

14. If P is any point on the circumference of a circle, centre C, which rolls with angular velocity ω on a fixed circle of centre O and radius a, prove that the angular velocity of P about O is

$$\omega - \frac{a\omega}{OP} \cos COP. \qquad \text{[Coll. Exam. 1912]}$$

15. A particle moves in the curve $y = a \log \sec \dfrac{x}{a}$ in such a way that the tangent to the curve rotates uniformly ; prove that the resultant acceleration of the particle varies as the square of the radius of curvature.

[S. 1925]

16. A circle and a tangent to it are given. A rod moves so that it touches the circle and one end is upon the tangent. Shew that the loci of the instantaneous centre in space and relative to the rod are both parabolas.

[S. 1925]

17. Shew that if two given points of a lamina describe coplanar straight lines, any point on a certain circle fixed in the lamina will also describe a straight line.   [S. 1917]

18. Prove that the motion of a rigid lamina moving in its own plane is at any instant (in general) equivalent to a rotation about a certain point I. What is the exceptional case ?

Prove that, if the vectors Oa, Ob represent the velocities of two points A, B, the triangles Oab, IAB are directly similar and that their corresponding sides are perpendicular.

Given Oa, Ob find a geometrical construction for the vector Oc which represents the velocity of a third point C. Shew in particular that $AC : CB = ac : cb$, and that, if ABC is a straight line, so is abc.

Four rods are freely jointed together so as to form a quadrilateral $PQRS$. Shew that if PQ is fixed the angular velocities of QR, PS are in the ratio $PT : QT$, where T is the point of intersection of PQ, RS.   [S. 1923]

19. If A and B are points on a rod which is moving in any way in a plane, and if Oa and Ob represent the velocities of A and B at any instant, prove that ab is perpendicular to AB. If C is any other point on the rod and if c divides ab in the same ratio as that in which C divides AB, prove that Oc represents the velocity of C at the same instant.

PQ, QR, RS are three rods in a plane jointed together at Q and R, and with the ends P and S jointed to fixed supports. If a triangle Oqr is drawn with Oq, qr, rO perpendicular to PQ, QR, RS respectively for any position of the rods, prove that as the rods move through this position Oq and Or represent on the same scale the velocities of Q and R.   [S. 1915]

ANSWERS

1. $22 \cdot 2$ hours.   2. $34° \, 49'$ E. of N.; 3 h. $5\frac{3}{4}$ m.; $31 \cdot 44$ m. N. of B.
3. At inclination $\cos^{-1}(ux/av)$ to the bank, where x is distance from the nearer bank. $\dfrac{2a}{u} \sin^{-1}\dfrac{u}{v}$.   4. E. or W. $185 \cdot 4$ m.   7. $u/2a$, $u/2a$, u/a ; 0.

12. If B, A are the centres of the larger and smaller rings and P any point on the latter, the vel. of P rel. to A is $a\omega$, of A rel. to B is $b\Omega - a\omega$, and of B is $b\Omega$. The point required is the end nearer to O of the diameter of the smaller circle parallel to OB.

Chapter VI

DYNAMICAL PROBLEMS IN TWO DIMENSIONS

6·1. In the early part of Chapter IV we interpreted Newton's law that *rate of change of momentum is proportional to the impressed force and takes place in the direction in which the force is impressed* as implying the equivalence of two vectors, the 'force' acting on a particle, and the product 'mass × acceleration'; and the rest of Chapter IV consists for the most part of examples of this equivalence in the case of rectilinear motion.

In the present Chapter we shall consider examples of this equivalence when a particle is free to move under the action of forces in one plane; the equivalence implying that the resolved parts of the two vectors in any assigned direction are equal. Consequently, if m be the mass of the particle, x, y its coordinates, and X, Y the sums of the resolved parts parallel to rectangular axes of all the forces acting upon the particle, we have the equations

$$m\ddot{x} = X \quad \text{and} \quad m\ddot{y} = Y.$$

In a large class of problems in dynamics of a particle the force components X and Y are given, and the solution of the problem consists in integrating these equations in order to determine the path of the particle.

6·2. Motion of Projectiles. Consider the case of a particle of mass m freely projected under the action of gravity in a non-resisting medium. Take for axes Ox, Oy the horizontal and upward vertical lines through the point of projection O, and let the particle be projected 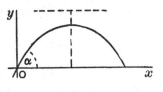 with velocity V in a direction making an angle α with the horizontal.

We have $\qquad\qquad X = 0$ and $Y = -mg$.

Therefore $\qquad\qquad \ddot{x} = 0$ and $\ddot{y} = -g$.

Integrate and introduce the initial values of $\dot{x}$, $\dot{y}$, namely $V\cos\alpha$, $V\sin\alpha$, and we get

$$\dot{x} = V\cos\alpha \text{ and } \dot{y} = V\sin\alpha - gt$$

therefore $\qquad x = Vt\cos\alpha \text{ and } y = Vt\sin\alpha - \tfrac{1}{2}gt^2 \quad \ldots\ldots\ldots(1).$

Eliminate t and we obtain the equation of the path of the particle

$$y = x\tan\alpha - \frac{1}{2}\frac{gx^2}{V^2}\sec^2\alpha \quad \ldots\ldots\ldots\ldots(2).$$

This represents a parabola and by writing the equation in the form

$$\left(x - \frac{V^2}{g}\sin\alpha\cos\alpha\right)^2 = -\frac{2V^2\cos^2\alpha}{g}\left(y - \frac{V^2\sin^2\alpha}{2g}\right),$$

it is seen that the latus rectum is $2V^2\cos^2\alpha/g$, the vertex the point $\left(\dfrac{V^2}{g}\sin\alpha\cos\alpha,\ \dfrac{V^2\sin^2\alpha}{2g}\right)$ and the axis vertically downwards.

The directrix is therefore horizontal, and its height above the point of projection is equal to the height of the vertex plus one-fourth of the latus rectum, i.e.

$$\frac{V^2\sin^2\alpha}{2g} + \frac{V^2\cos^2\alpha}{2g} \text{ or } \frac{V^2}{2g}.$$

Again we can shew that the velocity at any point of the path is, in magnitude, the velocity that would be acquired in falling freely from the directrix. For, if v is the velocity at time t,

$$v^2 = \dot{x}^2 + \dot{y}^2$$
$$= V^2 - 2Vgt\sin\alpha + g^2t^2$$
$$= 2g\left\{\frac{V^2}{2g} - (Vt\sin\alpha - \tfrac{1}{2}gt^2)\right\}$$
$$= 2g\left\{\frac{V^2}{2g} - y\right\}$$
$$= 2g \times \text{depth below the directrix.}$$

The *time of flight* before the projectile again reaches the horizontal plane through the point of projection is got by putting $y = 0$ in (1), which gives $t = 2V\sin\alpha/g$.

The *range* of the projectile on the horizontal plane through the point of projection is obtained as the value of x in (1) when for t we substitute the time of flight, i.e. $V^2 \sin 2\alpha/g$; and the greatest range for a given V is therefore V^2/g got by taking $\alpha = \tfrac{1}{4}\pi$.

Since $\sin 2\,(\tfrac{1}{2}\pi - \alpha) = \sin 2\alpha$, therefore in general there are two directions of projection with a given velocity which give the same horizontal range, viz. those of inclination α and $\tfrac{1}{2}\pi - \alpha$, the two coinciding when $\alpha = \tfrac{1}{4}\pi$.

6·21. *The range on an inclined plane through the point of projection* may be found from the equation of the path **6·2** (2) by writing $y = x \tan \beta$, where β is the inclination of the plane to the horizontal, i.e.

$$x \tan \beta = x \tan \alpha - \frac{1}{2}\frac{g x^2}{V^2} \sec^2 \alpha,$$

therefore
$$x = \frac{2V^2 \cos^2 \alpha}{g}\,(\tan \alpha - \tan \beta),$$

and the range required

$$= x \sec \beta$$
$$= \frac{2V^2}{g} \cos \alpha \sin (\alpha - \beta) \sec^2 \beta$$
$$= \frac{V^2}{g} \sec^2 \beta \,\{\sin (2\alpha - \beta) - \sin \beta\}.$$

For given values of V and β this expression is greatest when $2\alpha - \beta = \tfrac{1}{2}\pi$, and this makes the maximum range
$$r = V^2/g\,(1 + \sin \beta).$$

We may write the last result
$$\frac{V^2}{gr} = 1 + \sin \beta,$$

and we know that the polar equation of a parabola with the focus as pole is
$$\frac{l}{r} = 1 + \cos \theta,$$

where θ is measured from the vertex and l is the semi-latus rectum.

A comparison of the last two equations shews that, if we construct a parabola with focus at O, latus rectum $2V^2/g$, axis vertical and vertex A upwards, then any radius OP drawn from O will give the maximum range in direction OP for a particle

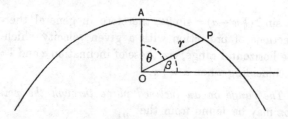

projected from O with velocity V. This parabola therefore is an outer boundary to the region that can be reached by projectiles starting from O with velocity V.

6·22. Geometrical Construction. A particle is to be projected from a point P so as to pass through a point Q. It may be shewn that with a given velocity of projection there may be two possible paths, one possible path or no possible path.

Thus if V be the given velocity, from P draw PM vertically upwards and of length $V^2/2g$, then M is a point on the directrix. The directrix of the path is therefore a horizontal line MN. Draw QN at right angles to the directrix. Then from the focus-directrix property of the parabola, the focus must lie on a

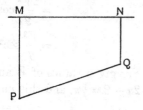

circle of centre P and radius PM, and also on a circle of centre Q and radius QN. If the circles intersect in two points S, S' either of them is the focus of a possible path and the bisector of the angle MPS, or MPS' gives the required direction of projection. Alternatively the circles may touch at a point S on PQ, which is then the focus of the one path possible; or the circles may not meet one another, in which case the velocity of projection is inadequate. The least velocity that will carry the particle from P to Q will correspond to the lowest possible position of the directrix, i.e. the position which makes

$$PM + QN = PQ.$$

Thus if v be the least velocity and $PQ = h$, and k be the height of Q above P, we have

$$PM + QN = h,$$
and
$$PM - QN = k,$$
so that
$$2PM = h + k,$$
and
$$v^2 = 2gPM = g\,(h + k).$$

6·23. Examples. (i) *Shew that the product of the two times of flight from P to Q with a given velocity of projection is $2PQ/g$.* [M. T. 1916]

Let V be the velocity of projection, α the inclination to the horizontal of the direction of projection, and a, b the horizontal and vertical distance of Q from P.

Then
$$a = Vt \cos \alpha,$$
and
$$b = Vt \sin \alpha - \tfrac{1}{2}gt^2 \,;$$
therefore
$$a^2 + (b + \tfrac{1}{2}gt^2)^2 = V^2 t^2,$$
or
$$\tfrac{1}{4}g^2 t^4 + (gb - V^2)\,t^2 + a^2 + b^2 = 0.$$

Hence, if t_1, t_2 are the two times of flight,
$$t_1^2 t_2^2 = 4\,(a^2 + b^2)/g^2 = 4PQ^2/g^2,$$
so that
$$t_1 t_2 = 2PQ/g.$$

(ii) *A gun fires a shell with a muzzle velocity 1040 feet per second. Neglecting the resistance of the air, what is the farthest horizontal distance at which an aeroplane at a height of 2500 feet can be hit and what gun elevation is required? Shew that the shell would then take approximately 44·2 seconds to reach the aeroplane.* [S. 1926]

This problem can be solved by regarding the equation of the path **6·2** (2) as giving the range x as a function of α, when for y the height of the aeroplane is substituted. We may then obtain another relation between x and α by differentiating with regard to α and using the fact that for a maximum value of x the derivative $dx/d\alpha$ must vanish. We can then solve the two equations for x and α, and also find the time taken from the relation $Vt \cos \alpha = x$.

A simpler solution is obtained from the consideration that the quadratic equation for t^2
$$\tfrac{1}{4}g^2 t^4 + (gb - V^2)\,t^2 + a^2 + b^2 = 0 \quad\ldots\ldots\ldots\ldots(1),$$
obtained in the last example must have real roots. In this case b denotes the height of the aeroplane and a denotes its horizontal distance. The condition for real roots is
$$(gb - V^2)^2 - g^2\,(a^2 + b^2) \not< 0.$$

The greatest value of a is therefore $\dfrac{V}{g}\,(V^2 - 2gb)^{\frac{1}{2}}$, and, taking $V = 1040$ and $b = 2500$, this gives $a = 31200$ ft.

Also for this value of a the quadratic (1) has equal roots in t^2, namely
$$t^2 = 2(V^2 - gb)/g^2;$$
and for the given values of V and b this makes $t = 44\cdot 2$ secs.

Reverting now to the equation for the horizontal range
$$a = Vt \cos \alpha,$$
we find that $\qquad\qquad \sec \alpha = 1\cdot 473,$
so that the elevation required is $47° 17'$.

6·3. Resisting Media. So far we have assumed that gravity is the only force acting on a projectile. Now suppose that the motion is opposed by a force proportional to the velocity. Thus if m denote the mass and V the velocity, let mkV denote the magnitude of the resistance. Therefore the components of the resistance parallel to horizontal and vertical axes Ox, Oy are
$$- mk\dot{x}, \quad - mk\dot{y}.$$

Let u, v denote the initial horizontal and vertical components of velocity. The equations of motion give
$$\ddot{x} = - k\dot{x} \quad \text{or} \quad \ddot{x} + k\dot{x} = 0,$$
therefore $\dot{x} + kx = u$, since initially $x = 0$ and $\dot{x} = u$.

Now multiply by the integrating factor e^{kt} and integrate both sides as in **1·61**,
$$xe^{kt} = \frac{u}{k} e^{kt} + C;$$
but $x = 0$ when $t = 0$, therefore $C = - u/k$; and, dividing by e^{kt},
$$x = \frac{u}{k}(1 - e^{-kt})\dots\dots\dots\dots\dots\dots\dots(1).$$

Again $\qquad\qquad \ddot{y} = - g - k\dot{y},$
therefore $\dot{y} + ky = v - gt$, since initially $y = 0$ and $\dot{y} = v$.

Now multiply by the integrating factor e^{kt} and integrate both sides
$$ye^{kt} = \int e^{kt}(v - gt) + C'$$
$$= \frac{v}{k} e^{kt} - \frac{g}{k} te^{kt} + \frac{g}{k^2} e^{kt} + C';$$
but $y = 0$ when $t = 0$, therefore $C' = - \frac{v}{k} - \frac{g}{k^2}$; and, dividing by e^{kt},
$$y = \left(\frac{v}{k} + \frac{g}{k^2}\right)(1 - e^{-kt}) - \frac{gt}{k} \dots\dots\dots\dots\dots(2).$$

Results (1) and (2) express x and y in terms of t, and if t is eliminated the resulting equation gives the path.

From (1) it follows that, as t becomes large, x tends to a constant value u/k, so that the path approaches a vertical asymptote.

6·31. Example. *A particle subject to gravity is projected at an angle α with the horizontal in a medium which produces a retardation equal to k times the velocity. It strikes the horizontal plane through the point of projection at an angle ω, and the time of flight is T. Prove that*

$$\frac{\tan \omega}{\tan \alpha} = \frac{e^{kT} - 1 - kT}{e^{-kT} - 1 + kT},$$

and deduce that $\omega > \alpha$.   [M. T. 1924]

Taking $u = V \cos \alpha$ and $v = V \sin \alpha$ as the initial components of velocity in the equations of the last article, the time of flight T is obtained by putting $y = 0$ in (2); i.e.

$$0 = \left(V \sin \alpha + \frac{g}{k} \right) (1 - e^{-kT}) - gT \quad\text{...............(3).}$$

Again $\tan \omega$ is the value of $-dy/dx$, when $y = 0$ and $t = T$, so that we have

$$\tan \omega = -\frac{\dot{y}}{\dot{x}} = -\frac{V \sin \alpha - gT}{V \cos \alpha - kx} = -\frac{V \sin \alpha - gT}{V \cos \alpha - V \cos \alpha (1 - e^{-kT})}$$

$$= -\frac{V \sin \alpha - gT}{V \cos \alpha e^{-kT}}.$$

Eliminate V by means of (3), first multiplying numerator and denominator by $(1 - e^{-kT})$, and we get

$$\tan \omega = \frac{\dfrac{g}{k} (1 - e^{-kT}) - gTe^{-kT}}{e^{-kT} \left\{ gT - \dfrac{g}{k} (1 - e^{-kT}) \right\} \cot \alpha},$$

therefore

$$\frac{\tan \omega}{\tan \alpha} = \frac{1 - e^{-kT} - kTe^{-kT}}{e^{-kT} (kT - 1 + e^{-kT})}$$

$$= \frac{e^{kT} - 1 - kT}{e^{-kT} - 1 + kT}.$$

If we expand e^{kT} and e^{-kT}, the numerator becomes

$$\frac{k^2 T^2}{2!} + \frac{k^3 T^3}{3!} + \dots,$$

a series of positive terms, and the denominator becomes

$$\frac{k^2 T^2}{2!} - \frac{k^3 T^3}{3!} + \dots,$$

the same series, save that the terms are alternately positive and negative. Therefore the numerator is greater than the denominator and $\omega > \alpha$.

6·32. Resistance proportional to the Square of the Velocity. We shall illustrate another law of resistance by taking an example in which the resistance varies as the square of the velocity.

A particle is projected vertically upwards under gravity. The resistance of the air produces an acceleration opposite to the velocity and numerically equal to kv^2, where v is the velocity and k a constant. If the initial velocity is V, and the square of kV^2/g can be neglected, shew that the particle reaches its highest point in time $\dfrac{V}{g} - \dfrac{kV^3}{3g^2}$, *and that the greatest altitude reached is*

$$\frac{V^2}{2g} - \frac{kV^4}{4g^2}.$$

If the initial velocity, in addition to the vertical component V, has a small horizontal component U, and the resistance follows the same law, shew that when the particle returns to the original level its horizontal velocity is approximately $Ue^{-kV^2/g}$. [M. T. 1925]

For the upward vertical motion, measuring y vertically upwards, we have

$$v\frac{dv}{dy} = -g - kv^2, \text{ or } \frac{dv^2}{dy} = -2g - 2kv^2.$$

Therefore
$$\frac{d(2kv^2)}{2kv^2 + 2g} = -2k\,dy\ ;$$

and, by integration,
$$\log(2kv^2 + 2g) = -2ky + C \quad \dotfill (1),$$

where C is a constant. Putting $v = V$ when $y = 0$ gives

$$C = \log(2kV^2 + 2g),$$

and substituting this value for C in (1),

$$\log\frac{kv^2 + g}{kV^2 + g} = -2ky.$$

The highest point is reached when $v = 0$, and then

$$2ky = \log\left(1 + \frac{kV^2}{g}\right)$$
$$= \frac{kV^2}{g} - \frac{k^2V^4}{2g^2} + \text{higher powers}\ ;$$

therefore the greatest altitude is $\dfrac{V^2}{2g} - \dfrac{kV^4}{4g^2}$.

To find the time, we write

$$\frac{dv}{dt} = -g - kv^2,$$

or
$$\frac{dv}{g + kv^2} = -dt.$$

Therefore $\dfrac{1}{\sqrt{kg}}\tan^{-1}\sqrt{\dfrac{k}{g}}\,v = C' - t$, where C' is a constant.

But when $t=0$ we have $v=V$, therefore $C' = \dfrac{1}{\sqrt{kg}} \tan^{-1} \sqrt{\dfrac{k}{g}}\, V.$

Hence $\dfrac{1}{\sqrt{kg}} \tan^{-1} \sqrt{\dfrac{k}{g}}\, v = \dfrac{1}{\sqrt{kg}} \tan^{-1} \sqrt{\dfrac{k}{g}}\, V - t.$

The time of reaching the highest point is found by putting $v=0$, and then

$$t = \frac{1}{\sqrt{kg}} \tan^{-1} \sqrt{\frac{k}{g}}\, V.$$

The function on the right-hand side can be expanded by Gregory's series $[\tan^{-1} x = x - \tfrac{1}{3}x^3 + \tfrac{1}{5}x^5 + \dots$, provided $-1 < x \leqslant 1]$, so that

$$t = \frac{1}{\sqrt{kg}} \left\{ \sqrt{\frac{k}{g}}\, V - \frac{1}{3}\left(\frac{k}{g}\right)^{\frac{3}{2}} V^3 + \dots \right\}$$

$$= \frac{V}{g} - \frac{kV^3}{3g^2}, \text{ neglecting higher powers.}$$

When there is also a horizontal velocity, we have for the horizontal acceleration

$$\ddot{x} = -kv^2 \cos\psi,$$

where ψ is the inclination of the path to the horizontal; but $v = \dot{s}$ and

$$\cos\psi = dx/ds,$$

therefore $\ddot{x} = -k\dot{s}^2\, dx/ds = -k\dot{s}\dot{x},$

or $\dfrac{\ddot{x}}{\dot{x}} = -k\dot{s}.$

Hence, by integration, $\log \dot{x} = C'' - ks$, and $\dot{x} = U$ when $s=0$, so that

$$C'' = \log U.$$

Therefore $\dot{x} = Ue^{-ks}.$

Now the horizontal velocity is small and decreasing, therefore the horizontal distance travelled is small, and as an approximation to the value of s when the particle again reaches the plane we may take twice the greatest altitude and write

$$ks = \frac{kV^2}{g} - \frac{k^2V^4}{2g^2}.$$

But by hypothesis the last term is negligible, so that $ks = kV^2/g$, and the required value of $\dot{x}$ is $Ue^{-kV^2/g}.$

6·4. Principle of Work. Reverting to the equations of motion of **6·1**, viz.

$$m\ddot{x} = X, \quad m\ddot{y} = Y,$$

if we multiply these equations respectively by $\dot{x}$ and $\dot{y}$ and add, we get

$$m(\dot{x}\ddot{x} + \dot{y}\ddot{y}) = X\dot{x} + Y\dot{y}.$$

Now integrate this equation with regard to t for any interval of time from t_0 to t_1, and it follows that

$$\tfrac{1}{2}m\,(\dot{x}^2+\dot{y}^2)_1 - \tfrac{1}{2}m\,(\dot{x}^2+\dot{y}^2)_0 = \int(X\,dx + Y\,dy),$$

or
$$\tfrac{1}{2}mv_1{}^2 - \tfrac{1}{2}mv_0{}^2 = \int(X\,dx + Y\,dy),$$

where the suffixes denote the values at the beginning and end of the interval. This result shews that *the increase in the kinetic energy of the particle in any interval is equal to the work done by the forces acting on the particle in that interval.*

Alternative proof. Let F be the resultant force acting on the particle when it is at any point P of its path, and ϵ the angle that the force makes with the tangent to the path.

Resolving along the tangent, we have

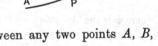

$$mv\,\frac{dv}{ds} = F\cos\epsilon.$$

Integrating along the path between any two points A, B, we get

$$\tfrac{1}{2}mv_1{}^2 - \tfrac{1}{2}mv_0{}^2 = \int F\cos\epsilon\,ds,$$

where v_0, v_1 are the velocities at A and B, and by **4·3** the integral represents the work done by the resultant force.

EXAMPLES

1. Shew that if a gun be situated on an inclined plane, the maximum range in a direction at right angles to the line of greatest slope is a harmonic mean between the maximum ranges up and down the plane respectively.   [S. 1910]

2. P, Q are two points distant a apart, and at heights h, k above a given horizontal plane; prove that the minimum velocity with which a particle must be projected from the plane so as to pass through P and Q is $\sqrt{\{g\,(a+h+k)\}}$.   [S. 1909]

3. A particle is projected with given velocity from a point P so as to pass through a point Q. If S is the focus of either of the possible trajectories, shew that the times of flight in the two trajectories are

$$\{(SP+SQ+PQ)^{\frac{1}{2}} \pm (SP+SQ-PQ)^{\frac{1}{2}}\}/g^{\frac{1}{2}}.$$

[S. 1917]

4. Prove that the velocity required to project a particle from a height h to fall at a horizontal distance a from the point of projection, is at least equal to $\sqrt{[g\{\sqrt{(a^2+h^2)}-h\}]}$.

The hammer was thrown at Cambridge a distance of 122 ft. in 1911; if the head of the hammer was 8 ft. from the ground at the instant of projection, calculate the least velocity required and the corresponding angle of projection.　[S. 1912]

5. A particle is projected under gravity from A so as to pass through B. Shew that if B has horizontal and vertical coordinates x, y referred to A, and the velocity of projection is $\sqrt{2gh}$, the angle between the two paths at B is a right angle if B lies on the ellipse $x^2+2y^2=2hy$.　[S. 1917]

6. Three particles are projected simultaneously and in the same vertical plane from a point with velocities v_1, v_2, v_3 in directions making angles α_1, α_2, α_3 with the horizontal. Shew that the area of the triangle formed by the particles at any time is proportional to the square of the time elapsed from the instant of projection, and that the three particles will always lie on a straight line, if $\Sigma v_2 v_3 \sin(\alpha_2-\alpha_3)=0$.　[S. 1912]

7. A particle P is describing a parabola freely under gravity. Shew that the angular velocity of the line joining P to the focus is $2gu/v^2$, where v is the velocity of the particle at P and u is the horizontal component of v.　[S. 1923]

8. A particle is projected from a given point with a velocity whose vertical component is given. Prove that the initial angular velocity about the focus of the path is greatest when the angle of projection is 45°.　[S. 1910]

9. Particles are projected simultaneously from a point under gravity in various directions with velocity V. Prove that at any subsequent time t they will all lie on a sphere of radius Vt, and determine the motion of the centre of this sphere.　[S. 1927]

10. A projectile is fired from a point O with a velocity due to a fall of 100 feet from rest, and hits a mark at a depth of 50 feet below O and at a distance of 100 feet from the vertical line through O. Shew that the two possible directions of projection are at right angles, and find to the nearest minute their inclinations to the horizontal.　[M. T. 1915]

11. A particle is projected at time $t=0$ in a fixed vertical plane from a given point S with given velocity $\sqrt{(2ga)}$, of which the *upward* vertical component is v. Shew that at time $t=2a/v$ the particle is on a fixed parabola (independent of v), that its path touches the parabola, and that its direction of motion is then perpendicular to its direction of projection.　[M. T. 1922]

12. A particle is projected from a point A so as to pass through a given point B and to return to the same level as A at a point C. Shew that the velocity with which the particle must be projected is

$$\sqrt{\{g\,(AN^2.NC^2+AC^2.NB^2)/2AN.NB.NC\}},$$

where N lies in AC directly below B.

13. Shew that all points in a vertical plane, which can be reached by shots fired with velocity v from a fixed point at a distance c from the plane, lie within a parabola of latus rectum $\dfrac{2v^2}{g}$, whose focus is at a distance $\dfrac{c^2g}{2v^2}$ vertically below the foot of the perpendicular on the plane from the point of projection.   [S. 1925]

14. A fort is on the edge of a cliff of height h. Shew that there is an annular region in which the fort is out of range of the ship, but the ship is not out of range of the fort, of area $8\pi kh$, where $\sqrt{2gk}$ is the velocity of the shells used by both.   [S. 1926]

15. A fort and a ship are both armed with guns which give their projectiles a muzzle velocity $\sqrt{2gk}$, and the guns in the fort are at a height h above the guns in the ship. If d_1 and d_2 are the greatest (horizontal) ranges at which the fort and ship, respectively, can engage, prove that

$$\frac{d_1}{d_2}=\sqrt{\frac{k+h}{k-h}}.$$ [M. T. 1928]

16. From a gun placed on a horizontal plane, which can fire a shell with velocity $\sqrt{2gH}$, it is required to throw a shell over a wall of height h, and the elevation of the gun cannot exceed α, where $\alpha < 45°$. Shew that this will be possible only when $h < H\sin^2\alpha$, and that, if this condition be satisfied, the gun must be fired from within a strip of the plane whose breadth is $4\cos\alpha\sqrt{H(H\sin^2\alpha-h)}$.   [S. 1911]

17. A shell fired with velocity V at elevation θ hits an airship at height H which is moving horizontally away from the gun with velocity v. Shew that if

$$(2V\cos\theta-v)\,(V^2\sin^2\theta-2gH)^{\frac{1}{2}}=vV\sin\theta,$$

the shell might also have hit the airship if the latter had remained stationary in the position it occupied when the gun was actually fired.
                    [M. T. 1917]

18. A battleship is steaming ahead with velocity V. A gun is mounted on the battleship so as to point straight backwards, and is set at an angle of elevation α. If v is the velocity of projection (relative to the gun), shew that the range is $\dfrac{2v}{g}\sin\alpha\,(v\cos\alpha-V)$; also shew that the angle of elevation for maximum range is $\cos^{-1}\{(V+\sqrt{V^2+8v^2})/4v\}$.   [S. 1925]

19. An aeroplane is flying with constant velocity v and at constant height h. Shew that, if a gun is fired point blank at the aeroplane after it has passed directly over the gun and when its angle of elevation as seen from the gun is α, the shell will hit the aeroplane provided

$$2\,(V\cos\alpha - v)\,v\tan^2\alpha = gh,$$

where V is the initial velocity of the shot, the path being assumed to be parabolic. [S. 1915]

20. The range of a rifle bullet is 1200 yards when α is the elevation of projection. Shew that if the rifle is fired with the same elevation from a car travelling at 10 miles per hour towards the target the range will be increased by $220\tan^{\frac{1}{2}}\alpha$ feet. [S. 1917]

21. A shot is fired with initial velocity V at a mark in the same horizontal plane; shew that if a small error $\epsilon°$ is made in the angle of elevation, and an error $2\epsilon°$ in azimuth, the shot will strike the ground at a distance from the mark $\dfrac{\pi V^2\epsilon}{90g}$.

Shew also that if the angle of elevation is less than about $31\frac{1}{2}°$ an error in elevation will cause the shot to miss the mark by a greater amount than an equal error in azimuth. [S. 1923]

22. A shell bursts on contact with the ground and pieces from it fly in all directions with all velocities up to 80 feet per second. Shew that a man 100 feet away is in danger for $\frac{5}{2}\sqrt{2}$ seconds. [S. 1921]

23. A particle moving under gravity describes a parabola of vertex A and focus S, the velocity at A being u. Prove that, when the particle is at P, where $SP = r$, the components of its velocity along and perpendicular to SP are respectively equal to the vertical and horizontal components of its velocity.

Shew that the component in the direction SP of the acceleration of the particle is $g - u^2/r$. [M. T. 1923]

24. An aeroplane travelling at a speed V relative to the air experiences a resistance $R = aV^2 + b/V^2$, where a and b are constants within certain limits of V. Shew that, within these limits of V, the power absorbed in air resistance has a minimum value H_0, at a speed V_0, where

$$H_0 = 4\,(ab^3/27)^{\frac{1}{4}}, \quad V_0^4 = b/3a.$$

Assuming that the effective thrust power of the propeller Z is independent of V, find the greatest rate of gain of height, and shew that the aeroplane is then climbing at an angle $\sin^{-1}\dfrac{Z - H_0}{WV_0}$ to the horizontal, where W is the weight of the aeroplane. [M. T. 1925]

25. A particle is projected horizontally with a velocity v_0 in a medium in which the resistance to its motion is kv^2 per unit mass, when its velocity is v. If ψ is the downward inclination of its path to the horizontal when it has traversed an arc s, shew, by resolving along the horizontal, that

$$v \cos \psi = v_0 e^{-ks},$$

and by resolving along the normal to the path, that

$$e^{2ks} \frac{ds}{d\psi} = \frac{v_0^2}{g} \sec^3 \psi. \qquad \text{[M. T. 1926]}$$

26. A particle P moving along a horizontal straight line has retardation ku, where u is the velocity at time t. When $t=0$, the particle is at O and has velocity u_0. Shew that $u_0 - u$ is proportional to OP, and that the final value of OP is u_0/k.

A particle subject to gravity describes a curved path in a resisting medium which causes retardation $k \times$ velocity. Shew that the resultant acceleration f has a constant direction, and that $f = f_0 e^{-kt}$, where f_0 is the acceleration when $t=0$. [M. T. 1923]

ANSWERS

3. 60·4 f.s. ; 43° 7′ with horizontal.   10. 76° 43′ ; − 13° 17′.

Chapter VII

HARMONIC MOTION

7·1. When a particle moves in a straight line so that its acceleration is always directed towards a fixed point in the line and is proportional to the distance from that point its motion is called **Simple Harmonic Motion**.

Let O be the fixed point and P the position of the particle at time t. Let $OP = x$, then the acceleration may be denoted by μx in magnitude, where μ is constant. But the acceleration is directed towards O, therefore the acceleration in the positive direction of the axis of x is $-\mu x$, and this will be so whether the particle is on the right or left of O. Hence in all positions of the particle we have

$$\ddot{x} = -\mu x \dots\dots\dots\dots\dots\dots\dots(1).$$

The solution of this equation may be written down as in **1·7** (16) or, as in **3·6** (ii) (β), we may multiply both sides by $2\dot{x}$ and integrate, thus

$$2\dot{x}\ddot{x} = -2\mu x\dot{x},$$

therefore $\dot{x}^2 = C - \mu x^2$ where C is a constant, which may be determined if the velocity is known in any one position.

This relation is equivalent to

$$\frac{dx}{\sqrt{(C - \mu x^2)}} = \pm dt,$$

where the positive or negative sign must be taken according as x is increasing or decreasing with t.

On integration we get

$$\sin^{-1}\sqrt{\frac{\mu}{C}}\,x = \pm\sqrt{\mu}\,t + \alpha,$$

where α is a constant,

or

$$x = \sqrt{\frac{C}{\mu}}\sin(\pm\sqrt{\mu}\,t + \alpha).$$

The solution contains two arbitrary constants C and α and may be written in the simpler form

$$x = A \sin \sqrt{\mu}\, t + B \cos \sqrt{\mu}\, t \quad \ldots\ldots\ldots\ldots\ldots (2).$$

This is the general form of solution of equation (1) and therefore includes all special cases.

For example : (i) Let the particle start from rest at a distance a from O; i.e. when $t = 0$, $x = a$ and $\dot{x} = 0$.

Then, since $\quad \dot{x} = A \sqrt{\mu} \cos \sqrt{\mu}\, t - B \sqrt{\mu} \sin \sqrt{\mu}\, t \ldots\ldots\ldots\ldots (3),$

on putting $t = 0$ in (2) and (3) we get

$$a = B, \quad \text{and} \quad 0 = A,$$

therefore $\qquad x = a \cos \sqrt{\mu}\, t \quad \ldots\ldots\ldots\ldots\ldots\ldots (4)$

gives the position at time t in this case.

(ii) Let the particle be at a distance a from O and moving towards O with velocity v when $t = 0$.

Here we have $x = a$ and $\dot{x} = -v$ when $t = 0$. Therefore, from (2) and (3), $a = B$, and $-v = A \sqrt{\mu}$, so that in this case the position at time t is given by

$$x = -\frac{v}{\sqrt{\mu}} \sin \sqrt{\mu}\, t + a \cos \sqrt{\mu}\, t \quad \ldots\ldots\ldots\ldots (5).$$

7·11. Periodicity. The solutions obtained in the last article all represent periodic motions, for, if t is increased by $2\pi/\sqrt{\mu}$ or by any multiple of $2\pi/\sqrt{\mu}$, the values obtained for x and $\dot{x}$ in (2), (3), (4), (5) remain unaltered, shewing that after a time $2\pi/\sqrt{\mu}$ the particle is again in the same position and moving with the same velocity as before.

The interval $2\pi/\sqrt{\mu}$ is called the **period** of the harmonic motion or oscillation; and the **frequency** of the oscillations means the number of complete oscillations per second, i.e. $\sqrt{\mu}/2\pi$.

The distance through which the particle moves away from O on either side is called the **amplitude** of the oscillation. We notice that the period only depends on the given constant μ, so that the period is independent of the amplitude.

Further when a coordinate of position is represented by a formula $a \cos(nt + \epsilon)$ ϵ is called the **phase** at the instant from which t is measured. If two different harmonic motions with the same period $2\pi/n$ are represented by

$$a\cos(nt+\epsilon),\quad b\cos(nt+\epsilon'),$$

they are said to differ in phase. The difference in phase $\epsilon-\epsilon'$ may be spoken of as a certain number of radians, or it may be measured in terms of the time in which either of the angles concerned would increase by this amount, i.e. $(\epsilon-\epsilon')/n$ seconds, or it may be measured as the fraction $(\epsilon-\epsilon')/2\pi$ of a period. Thus if two harmonic motions are represented by

$$a\cos nt,\quad b\sin nt,$$

it follows that, since $\sin nt=\cos\left(nt-\dfrac{\pi}{2}\right)$, these motions differ in phase by one-quarter of a period, i.e. one particle is at the origin when the other is at an extreme position and vice versa.

7·2. Geometrical Representation. Let a point Q describe a circle of centre O with uniform angular velocity ω. Let P be the projection of Q on a fixed diameter AA'. As Q moves round the circle P will oscillate to and fro in the line AA', and we can shew that the motion of P is simple harmonic motion.

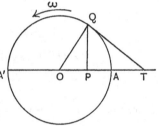

Since Q moves uniformly in a circle its only acceleration is ω^2QO along QO. The acceleration of P is the resolved part along AA' of the acceleration of Q, i.e. ω^2PO. Hence the acceleration of P is always directed towards O and proportional to its distance from O and therefore the motion of P is simple harmonic.

The period of oscillation of P is the time taken for Q to describe the circle, i.e. $2\pi/\omega$. This agrees with the $2\pi/\sqrt{\mu}$ of 7·11 because, by comparing the formulae for acceleration, we see that $\omega^2=\mu$.

The velocity of Q in the circle is ωQO along the tangent TQ. The velocity of P is the resolved part along AA' of the velocity of Q, i.e. $\omega QO\cos QTP=\omega QO\cos OQP=\omega PQ$ in magnitude. The velocity of P is therefore proportional to the ordinate PQ and has its greatest value when P is at O.

7·21. *A particle starts at a distance a from O with velocity v towards O and acceleration = μ × distance directed towards O; to find the uniform circular motion of which this motion is a projection.*

Take $OM = a$, and erect a perpendicular MN so that

$$\sqrt{\mu} \, MN = v.$$

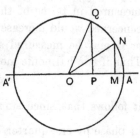

Join ON, and this will be the radius of the required circle. For if a point Q moves round the circle with angular velocity $\sqrt{\mu}$ then by **7·2** its projection P on OM has acceleration μPO towards O, and its velocity when P is at M is $\sqrt{\mu}\, MN$, which we have made equal to the given velocity v.

We can deduce formula (5) of **7·1** thus: Measuring t from the instant when the particle is at M, in time t OQ turns through the angle $NOQ = \sqrt{\mu}\,t$, and, if $MON = \epsilon$, we have

$$x = OP = OQ \cos (\sqrt{\mu}\,t + \epsilon)$$
$$= ON \cos \epsilon \cos \sqrt{\mu}\,t - ON \sin \epsilon \sin \sqrt{\mu}\,t$$
$$= OM \cos \sqrt{\mu}\,t - MN \sin \sqrt{\mu}\,t,$$

therefore $$x = a \cos \sqrt{\mu}\,t - \frac{v}{\sqrt{\mu}} \sin \sqrt{\mu}\,t.$$

7·22. The geometrical representation of simple harmonic motion is very useful in the solution of examples. The revolving radius may be regarded as a time-keeper of the motion, since the angle through which it turns between any two positions is proportional to the time taken.

7·23. It is to be noted that an equation of the form

$$\ddot{x} + \mu x = 0,$$

where μ is a positive number, always represents a periodic motion, of period $2\pi/\sqrt{\mu}$ independent of the amplitude.

7·3. Elastic Strings and Spiral Springs. Hooke's Law. The 'extension' of a stretched elastic string means the ratio of the increment in length to the unstretched length. Thus if l_0 is the natural or unstretched length and the stretched length is l then the extension is $(l - l_0)/l_0$*.

Hooke's Law is that the *tension of the string is proportional to the extension.* If T denote the tension and we state the law in the form

$$T = \lambda \frac{l - l_0}{l_0},$$

then λ is called the modulus of elasticity of the string.

The extension or compression of a spiral spring follows the same law, but in this case the length is measured along the axis of the helix and not along the wire that forms the spring; and when the spring is extended or compressed the force exerted by the spring is a tension or a thrust in the direction of the axis. The formula above may be used for compression as well as extension provided we regard a negative tension as a thrust. For when the spring is compressed the length l is less than the natural length l_0, so that the formula would give a negative tension, i.e. a thrust of magnitude $\lambda (l_0 - l)/l_0$.

7·31. Work done in Stretching an Elastic String. Let OA represent the natural length of a string. Suppose the end O to be fixed and the string stretched by moving the end A.

To find the work done as the length is increased from OB to OC.

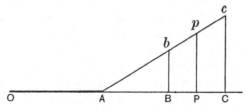

Let P be a point between B and C and at every such point erect an ordinate Pp to represent the tension. Since the tension is proportional to the extension, therefore the ratio $Pp : AP$ is constant. Hence the upper ends of the ordinates lie on a straight line $Abpc$; and this line is a 'force-space curve,' **4·43**.

---

* The above is the proper mathematical meaning of the word *extension*, it is however frequently used to denote *increment of length*.

In order to stretch the string the experimenter has to apply to
the end A a force equal to the tension; the line BPC is the
'space' through which the point of application of the gradually
increasing force is moved, and the area under the 'force-space
curve' represents the work done. Therefore the work done in
extending the string from a length OB to a length OC is
measured by the area

$$BCcb = \tfrac{1}{2}(Bb + Cc)\,BC$$

= increase in length × mean of initial and final tensions.

7·32. The same result may be obtained by using Hooke's
formula for tension. Thus if T be the tension when the length
is x, we take $T = \lambda(x - l_0)/l_0$.

Also the work done when a force T moves its point of application through a distance dx in its line of action is $T\,dx$; therefore
in increasing the length from l to l' the work done is

$$\int_l^{l'} T\,dx = \frac{\lambda}{l_0}\int_l^{l'}(x - l_0)\,dx$$

$$= \tfrac{1}{2}\frac{\lambda}{l_0}\{(l' - l_0)^2 - (l - l_0)^2\}$$

$$= \tfrac{1}{2}(l' - l)\left\{\frac{\lambda(l' - l_0)}{l_0} + \frac{\lambda(l - l_0)}{l_0}\right\}$$

= increase in length × mean of initial and final tensions.

7·4. The motion of a heavy particle suspended from a fixed
point by an elastic string or a spiral spring is simple harmonic
motion, because the tension of the string causes an acceleration
proportional to the distance from a fixed
point and directed towards it.

Let A be the fixed point and AB the
unstretched length of the string. Let m be
the mass of the particle and let the weight
of the particle be such as to extend the
string to the length AO when the particle
hangs at rest.

Let $AB = l_0$, $AO = l$ and measure x vertically downwards from O. In the equilibrium
position the weight is equal to the tension, so that

$$mg = \lambda(l - l_0)/l_0 \quad\dots\dots\dots\dots\dots(1).$$

Let the particle be in another position P where $OP = x$; it then has a downward acceleration given by

$$m\ddot{x} = \text{downward force}$$

$$= mg - T \dots\dots\dots\dots\dots\dots\dots(2);$$

where T is the tension when the length is $AP = l + x$, so that

$$T = \lambda\,(l + x - l_0)/l_0 \dots\dots\dots\dots\dots\dots(3).$$

Using (1) and (3), equation (2) becomes

$$m\ddot{x} = - \lambda x/l_0 \dots\dots\dots\dots\dots\dots\dots(4).$$

This equation represents a simple harmonic motion of period

$$2\pi \surd(ml_0/\lambda).$$

Up to this point it has not been necessary to specify in what way the motion began. Equation (4) is independent of initial conditions.

As a particular case let us suppose that the particle is pulled downwards to a point C, where $OC = a$ and then set free. If $OC \leqslant OB$ the equation (4) will apply to the whole motion, which will be a purely oscillatory motion about O as centre through a distance a below and above O. But if $OC > OB$, the string will become slack when the ascending particle passes B. Draw a circle of centre O and radius OC. Equation (4) can be written $\ddot{x} = - \omega^2 x$ where $\omega = \surd(\lambda/ml_0)$; and the simple harmonic motion is the projection on the vertical of the motion of a point describing the circle with angular velocity ω. This motion lasts until the particle reaches B, and the time from C to B

$$= \text{the angle } COD/\omega$$

$$= \frac{1}{\omega}\left(\pi - \cos^{-1}\frac{l - l_0}{a}\right).$$

At B the velocity is $v = \omega BD = \omega\,\surd\{a^2 - (l - l_0)^2\}$, and the particle then moves freely upwards under gravity, returning to B in time $2v/g$. Hence the whole time of ascent and descent

$$= \frac{2\omega}{g}\,\surd\{a^2 - (l - l_0)^2\} + \frac{2}{\omega}\left\{\pi - \cos^{-1}\frac{l - l_0}{a}\right\}$$

where $\omega = \surd(\lambda/ml_0)$.

7·5. Further Applications. In the working of examples on
elastic strings or springs it is not necessary that the data should
include the natural length and the modulus of elasticity provided that the ratio of an additional force to the additional
length it produces is given.

Example. (i) *A body of mass 5 lb. is hung on a light spring and is
found to stretch it 6 ins. The mass is then pulled down a further 2 ins. and
released. Find the period of the oscillations and the kinetic energy of the
mass as it passes through its equilibrium position.*

Since a stretch of ·5 ft. is due to a force of 5 lb. weight therefore a
stretch of x ft. would correspond to a tension of $10x$ lb. weight.

In the equilibrium position the weight is balanced by the tension, and
when the body is at a distance x ft. below the equilibrium position there
is therefore an extra tension of $10xg$ poundals acting upwards on the body;
and it is this extra unbalanced tension which causes acceleration, so that

$$5\ddot{x} = -10xg,$$

or $$\ddot{x} = -64x.$$

Hence the period $= 2\pi/\sqrt{64} = \tfrac{1}{4}\pi = ·785$ sec.

Again the velocity is greatest as the body passes through the equilibrium
position and is then $\omega \times$ amplitude, where $\omega^2 = 64$ and amplitude $= 2$ ins.
$= \tfrac{1}{6}$ ft.

Therefore the velocity $v = 4/3$ f.s., and the kinetic energy

$$\tfrac{1}{2}mv^2 = \tfrac{1}{2} \times 5 \times \tfrac{16}{9} \text{ absolute units} = \tfrac{40}{9} \times \tfrac{1}{32} = \tfrac{5}{36} \text{ foot pound.}$$

(ii) *A cage of mass M lb. is being pulled up with uniform velocity u by
a long steel cable when the upper end of the cable is suddenly fixed. Having
given that a weight of m lb. would extend the cable 1 ft., shew that the
amplitude of the oscillation of the cage is $u\sqrt{(M/mg)}$.*

If the cage were hanging from the cable at rest, the cable, though
extended by the weight of the cage, would have a definite length which
we may call the equilibrium length, and such a position of the cage may
be called an equilibrium position. When the upward motion is uniform,
since there is no acceleration, the length of the cable will remain the
equilibrium length and the position of the cage relative to the upper end
of the cable will still be the equilibrium position.

We assume that any vertical displacement of the cage which extends or
compresses the cable results in an extra tension or thrust proportional to
the extension or compression. Hence, after the fixing of the upper end of
the cable, the motion of the cage becomes a simple harmonic motion about
its equilibrium position, starting from this position with velocity u.

Since a weight of m lb. would extend the cable 1 ft., therefore a
displacement of the cage through x ft. from the equilibrium position

would be opposed by an unbalanced force of mx pounds weight, so
that

$$M\ddot{x} = -mgx,$$

or $$\ddot{x} = -\omega^2 x,$$

where $\omega^2 = mg/M$.

But if a is the amplitude of the oscillation the velocity at the centre of
the harmonic motion is $a\omega$.

Therefore $a\omega = u$; but $\omega = \sqrt{(mg/M)}$,

so that $a = u\sqrt{(M/mg)}$.

(iii) *A heavy particle is supported in equilibrium by two equal elastic
strings with their other ends attached to two points in a horizontal plane
and each inclined at an angle of 60° to the vertical. The modulus of
elasticity is such that when the particle is suspended from any portion of the
string its extension is equal to its natural length. The particle is displaced
vertically a small distance and then released. Prove that the period of its
small oscillations is $2\pi\sqrt{2l/5g}$, where l is the stretched length of either string
in equilibrium.*   [S. 1923]

Let m be the mass of the particle and λ the modulus of elasticity. Then
by supposing the particle to be
suspended from any portion of the
string, since the extended length is
double the natural length we find
that $\lambda = mg$.

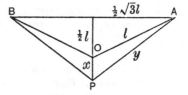

If l_0 be the natural length of
either string, we have, in the equilibrium position,

$$mg = 2\lambda\,\frac{l - l_0}{l_0}\cos 60° = \lambda\,\frac{l - l_0}{l_0}\,;$$

but $\lambda = mg$, therefore $l_0 = \tfrac{1}{2}l$.

Let x denote the vertical displacement and y the length of either string
at time t. To find the period of small oscillations we want to obtain an
equation of the form

$$\ddot{x} = -\mu x,$$

where μ is a constant. It will therefore be sufficient for our purpose to
write down the equation of motion at time t and neglect all powers of x
higher than the first.

We have $$m\ddot{x} = mg - 2\lambda\frac{y - l_0}{l_0}\cos OPA,$$

where P is the particle at time t, O is its equilibrium position and
$PA = PB = y$ are the strings.

Now
$$y^2 = (x + \tfrac{1}{2}l)^2 + \tfrac{3}{4}l^2$$
$$= l^2 + lx + x^2 \ ;$$

therefore
$$y = l\left(1 + \frac{x}{l}\right)^{\frac{1}{2}} = l + \tfrac{1}{2}x,$$

correct to the first power of x, and
$$\cos OPA = \frac{\tfrac{1}{2}l + x}{y} = \frac{\tfrac{1}{2}l + x}{l + \tfrac{1}{2}x}$$
$$= \frac{1}{2}\left(1 + \frac{2x}{l}\right)\left(1 - \frac{x}{2l}\right)$$
$$= \frac{1}{2}\left(1 + \frac{3x}{2l}\right),$$

to the first power of x.

Hence
$$m\ddot{x} = mg - \frac{2\lambda\,(l + \tfrac{1}{2}x - \tfrac{1}{2}l)}{\tfrac{1}{2}l} \cdot \frac{1}{2}\left(1 + \frac{3x}{2l}\right),$$

therefore
$$\ddot{x} = g - g\left(1 + \frac{x}{l}\right)\left(1 + \frac{3}{2}\frac{x}{l}\right),$$

or
$$\ddot{x} = -\frac{5}{2}\frac{gx}{l}\ ;$$

which represents a simple harmonic motion of period $2\pi\,\sqrt{2l/5g}$.

(iv) *A warship is firing at a target 3000 yards away dead on the beam, and is rolling (simple harmonic motion) through an angle of 3° on either side of the vertical in a complete period of 16 secs. A gun is fired during a roll 2 secs. after the ship passes the vertical. The gun was correctly aimed at the moment of firing, but the shell does not leave the barrel till 0·03 sec. later. Shew that the shell will miss the centre of the target by about 4 feet.*

[S. 1923]

Let θ denote the angle turned through by the ship in t seconds after passing the vertical. Then the change in θ is simple harmonic, so that it is connected with t by an equation
$$\ddot{\theta} = -n^2\theta \quad\dotfill(1),$$
where $2\pi/n =$ the complete period $= 16$ secs.

The complete solution of (1) is
$$\theta = A\sin nt + B\cos nt,$$
but θ vanishes for $t = 0$, therefore $B = 0$, and
$$\theta = A\sin nt.$$

Also A is the amplitude of the oscillation, i.e. an angle of 3° or $\pi/60$ radians.

The angular velocity is therefore given by
$$\dot{\theta} = nA\cos nt\ ;$$
and 2 secs. after passing the vertical the value of this is
$$\frac{\pi}{8}\cdot\frac{\pi}{60}\cos\frac{\pi}{4}\ ;$$

or taking π^2 as equal to 10, the angular velocity of the ship at the instant of firing the gun is $1/48\,\sqrt{2}$ radians per sec.

It follows that during the 0·03 sec. before the shell leaves the barrel the gun receives an additional angular elevation $\delta = \cdot03/48\ \sqrt{2}$ radians.

Now if V be the velocity and α the angular elevation of projection,

$$(V^2 \sin 2\alpha)/g = 9000 \text{ feet};$$

and, if Δ denote the additional range when the elevation is $\alpha + \delta$,

$$V^2 \sin 2\ (\alpha + \delta)/g = 9000 + \Delta;$$

or $\qquad\qquad V^2 (\sin 2\alpha + 2\delta \cos 2\alpha)/g = 9000 + \Delta,$

so that $\qquad\qquad\qquad \Delta = 18000\delta \cot 2\alpha.$

Also the shell will pass over the centre of the target at a height $\Delta \tan(\alpha + \delta)$ approximately; and if we neglect the square of the angular elevation, this is equal to 9000δ, and substituting the value found for δ this gives 4 feet as the approximate result.

7·6. Simple Pendulum. The simple pendulum consists of a heavy particle suspended from a fixed point by a fine thread moving in a vertical plane.

Let m be the mass of the particle and l the length of the thread. If at time t the thread makes an angle θ with the vertical the acceleration of the particle along the tangent to its path is $l\ddot{\theta}$. The forces acting on the particle are its weight mg and the tension of the thread and the former alone has a component along the tangent. Hence by resolving along the tangent we have

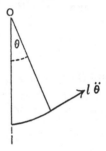

$$ml\ddot{\theta} = -mg \sin \theta.$$

If the oscillations are so small that we may put θ instead of $\sin \theta$, the equation becomes

$$l\ddot{\theta} = -g\theta.$$

This equation represents a harmonic oscillation of period

$$2\pi \sqrt{(l/g)}.$$

A " seconds pendulum " is one of which a single swing or half-period occupies one second, so that the length of the seconds pendulum is given by $\pi \sqrt{(l/g)} = 1$, or $l = g/\pi^2$.

7·61. Equivalent Simple Pendulum. Any simple harmonic motion may be compared with the motion of a simple pendulum and such motions may be regarded as equivalent if their periods are the same.

Thus, if we compare the equations

$$\ddot{x} = -\mu x \text{ and } l\ddot{\theta} = -g\theta,$$

they represent equivalent motions if $\mu = g/l$; and we may therefore say with regard to the motion $\ddot{x} = -\mu x$ that the 'length of the equivalent simple pendulum' is g/μ.

Example. *A particle of mass m is attached to the middle point of a string of length 2b which is tightly stretched between two fixed points. To*

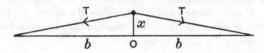

find the length of the equivalent simple pendulum, when the particle is displaced at right angles to the string, neglecting the force of gravity in comparison with the tension of the string.

If x denotes the displacement of the particle, the increase in length of the string is $2\{\sqrt{(b^2 + x^2)} - b\}$ and, if we neglect all powers of x above the first, this is zero. Therefore the tension T remains constant and by resolving at right angles to the string we get

$$m\ddot{x} = -2Tx/\sqrt{(b^2 + x^2)},$$

or, neglecting x^2,
$$m\ddot{x} = -2Tx/b.$$

Comparing this equation with $l\ddot{\theta} = -g\theta$, we see that the length of the equivalent simple pendulum is given by

$$\frac{l}{g} = \frac{mb}{2T},$$

or
$$l = mgb/2T.$$

7·62. As an illustration of the finite oscillations of a simple pendulum we will take the following:

Example. *A particle moving along the axis of x has an acceleration Xx towards the origin, where X is a positive function of x which is unchanged when −x is put for x. The periodic time, when the particle vibrates between x = −a and x = a, is T. Shew that*

$$2\pi/\sqrt{X_1} < T < 2\pi/\sqrt{X_2},$$

where X_1, X_2 are the greatest and least values of X within the range $x = -a$ to $x = a$.

Shew that, when a simple pendulum of length l vibrates through 30° on either side of the vertical, T lies between

$$2\pi\sqrt{l/g} \text{ and } 2\pi\sqrt{l/g} \times \sqrt{\pi/3}.$$

We have $\ddot{x} = -Xx$, throughout the motion, therefore

$$2\dot{x}\ddot{x} = -2Xx\dot{x}.$$

Integrate this, remembering that the velocity $\dot{x}$ vanishes when $x = a$, and we get

$$\dot{x}^2 = -2\int_a^x Xx\,dx = 2\int_x^a Xx\,dx.$$

Therefore $dt = \dfrac{-dx}{\sqrt{\left\{2\displaystyle\int_x^a Xx\,dx\right\}}}$, the negative sign since we have supposed

the motion to begin from $x = a$, so that x is decreasing as t increases. The period is four times the time from $x = a$ to $x = 0$, therefore

$$T = 4\int_0^a \frac{dx}{\sqrt{\left\{2\displaystyle\int_x^a Xx\,dx\right\}}}.$$

Now X is a positive function, and

$$X_1 > X > X_2,$$

therefore $\qquad 2\displaystyle\int_x^a X_1 x\,dx > 2\int_x^a Xx\,dx > 2\int_x^a X_2 x\,dx\,;$

and X_1, X_2 are constants, so that the last line gives

$$X_1(a^2 - x^2) > 2\int_x^a Xx\,dx > X_2(a^2 - x^2).$$

Hence

$$4\int_0^a \frac{dx}{\sqrt{\{X_1(a^2 - x^2)\}}} < 4\int_0^a \frac{dx}{\sqrt{\left\{2\displaystyle\int_x^a Xx\,dx\right\}}} < 4\int_0^a \frac{dx}{\sqrt{\{X_2(a^2 - x^2)\}}}\,;$$

but $\qquad 4\displaystyle\int_0^a \frac{dx}{\sqrt{(a^2 - x^2)}} = 4\left[\sin^{-1}\frac{x}{a}\right]_0^a = 2\pi,$

therefore $\qquad 2\pi/\sqrt{X_1} < T < 2\pi/\sqrt{X_2}.$

Again the equation of motion of the simple pendulum

$$l\ddot{\theta} = -g\sin\theta$$

may be written $\qquad \ddot{\theta} = -\dfrac{g}{l}\dfrac{\sin\theta}{\theta}\cdot\theta,$

and $\sin\theta/\theta$ is positive and unchanged when $-\theta$ is put for θ, so that the last result may be applied. Also as θ increases from 0 to $\pi/6$, $\sin\theta/\theta$ decreases from 1 to $3/\pi$, and therefore the greatest and least values of $\dfrac{g}{l}\dfrac{\sin\theta}{\theta}$ are $\dfrac{g}{l}$ and $\dfrac{g}{l}\dfrac{3}{\pi}$. Hence the period lies between

$$2\pi\sqrt{l/g} \text{ and } 2\pi\sqrt{l/g} \times \sqrt{\pi/3}.$$

7·7. Disturbed Simple Harmonic Motion. Suppose that in addition to the force varying as the distance there is a force in the line of motion producing a constant acceleration f in the positive direction of the axis of x. Then

$$\ddot{x} = -\mu x + f,$$

or

$$\ddot{x} = -\mu (x - f/\mu).$$

If we put $x = x' + f/\mu$, which is equivalent to moving the origin to the point of equilibrium $x = f/\mu$, the equation becomes

$$\ddot{x}' = -\mu x'.$$

This shews that the effect of the constant force is merely to move the centre of the simple harmonic motion a definite distance f/μ in the direction of the constant force; i.e. the centre of the simple harmonic motion is the point of equilibrium, or the point where the variable force is balanced by the constant force. We have had applications of this in problems of masses suspended by elastic strings or springs (**7·4, 7·5**) where the constant force of gravity causes the extension of the string or spring and the oscillations take place about an equilibrium position.

7·71. Periodic Disturbing Force. Forced Oscillations.

The equation $\qquad \ddot{x} = -\mu x + f \cos pt$ ....(1)

represents harmonic motion disturbed by a periodic force proportional to $\cos pt$.

To find a particular integral of this equation, substitute

$$x = C \cos pt$$

and we get $\qquad -Cp^2 \cos pt = -\mu C \cos pt + f \cos pt;$

so that $x = C \cos pt$ is a solution provided that

$$C = f/(\mu - p^2); \quad \text{i.e. } x = \frac{f \cos pt}{\mu - p^2}$$

is a particular integral. The complete solution is found by adding to the particular integral the complete solution of

$$\ddot{x} + \mu x = 0,$$

which by **1·7** (16) or by **7·1** (2) is

$$A \sin \sqrt{\mu}\,t + B \cos \sqrt{\mu}\,t,$$

so that $\qquad x = A \sin \sqrt{\mu}\,t + B \cos \sqrt{\mu}\,t + \dfrac{f \cos pt}{\mu - p^2}$ ....(2).

The two parts of the solution represent what are called the 'free' and the 'forced' oscillations, the latter being of the same period $2\pi/p$ as the disturbing force. Either free or forced oscillations can exist independently and the actual motion may be a combination of both, depending on the initial circumstances.

When p^2 is nearly equal to μ, i.e. when the forced and the free oscillations have nearly the same period, the amplitude $f/(\mu - p^2)$ of the forced oscillations is large. When $p^2 = \mu$, equation (1) no longer has a solution of the form $x = C \cos pt$; but if we substitute $x = Ct \sin pt$ in (1) we get

$$C(-p^2 t \sin pt + 2p \cos pt) = -\mu Ct \sin pt + f \cos pt,$$

which gives $C = f/2p$, so that the forced oscillation is now represented by $x = \dfrac{f}{2p} t \sin pt$; shewing that the amplitude of the forced oscillation increases continuously with t.

The phenomenon of the large amplitude of vibration of a body forced to vibrate with a period nearly equal to that of its free vibrations is a familiar one in the theory of sound and of electromagnetism and is known as *resonance*. The reason why troops are ordered to fall out of step in crossing a bridge is lest by forcing upon the structure vibrations of period nearly the same as that of a free vibration, displacements of large amplitude should be set up and cause the bridge to break.

7·72. Example. *Consider again the problem of* **7·61,** *supposing that one end of the string is fixed while the other end has a periodic motion of small amplitude at right angles to the string such that the displacement at time t is given by* $\xi = a \sin pt$.

As before, if x denotes the displacement of the particle, and we neglect x^2 and a^2, we can shew that the tension T remains constant. Resolving in the direction of x, we have, for the motion of the particle,

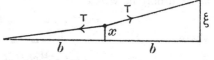

$$m\ddot{x} = -Tx/b + T(\xi - x)/b,$$

or, putting $2T/mb = n^2$, we get

$$\ddot{x} = -n^2 x + \tfrac{1}{2}n^2 a \sin pt.$$

The solution is, as in **7·71,**

$$x = A \sin nt + B \cos nt + \tfrac{1}{2}\,\frac{n^2 a \sin pt}{n^2 - p^2} \quad \dots\dots\dots\dots\dots(3).$$

Suppose that when $t = 0$, $x = 0$ and $\dot{x} = 0$; the first of these conditions requires that $B = 0$, and the second makes $0 = nA + \frac{1}{2} \frac{n^2 a p}{n^2 - p^2}$. In this case therefore the solution (3) takes the form

$$x = \frac{1}{2} \frac{na}{n^2 - p^2} (n \sin pt - p \sin nt),$$

and represents a combination of forced and free oscillations.

If however we suppose that when $t = 0$, $x = 0$ and that the velocity $\dot{x}$ is then $\frac{1}{2} \frac{n^2 a p}{n^2 - p^2}$, we find that both A and B are zero, so that with this set of initial conditions (3) becomes

$$x = \frac{1}{2} \frac{n^2 a \sin pt}{n^2 - p^2},$$

shewing that now there are no 'free' oscillations but the 'forced' oscillations alone.

7·8. Damped Harmonic Oscillations.

When a particle moves in a straight line under the action of a force towards a fixed point in the line, varying as the distance, and the motion is resisted by a force proportional to the velocity, we may write the equation of motion

$$\ddot{x} = - \mu x - 2k\dot{x},$$

or
$$\ddot{x} + 2k\dot{x} + \mu x = 0 \quad ....(1).$$

This equation is solved in **1·7**. There are three cases.

If $\mu < k^2$, the solution is of the form

$$x = A e^{-kt + \sqrt{(k^2 - \mu)} \, t} + B e^{-kt - \sqrt{(k^2 - \mu)} \, t} \quad ......\mathbf{1·7} (10),$$

and the motion is non-oscillatory.

If $\mu = k^2$, the solution is of the form

$$x = e^{-kt} (A + Bt) \quad ....\mathbf{1·7} (11),$$

which again does not represent oscillations; but if $\mu > k^2$, the solution may be written

$$x = e^{-kt} (A \cos nt + B \sin nt) \quad ....\mathbf{1·7} (12),$$

where
$$n^2 = \mu - k^2,$$

or
$$x = C e^{-kt} \cos (nt + \epsilon),$$

where C and ϵ are arbitrary constants.

This last result represents harmonic oscillations with amplitude Ce^{-kt} decreasing steadily with the time. They are called *damped oscillations* and e^{-kt} is the damping coefficient. The oscillations die away as t increases.

7·9. Damped Forced Oscillations. The equation

$$\ddot{x} + 2k\dot{x} + \mu x = f \cos pt$$

represents damped forced oscillations. The complete solution of this equation is got by adding a particular integral to the solution of equation (1) of **7·8**.

To find a particular integral we substitute

$$x = E \cos pt + F \sin pt$$

as a trial solution, and we find that it satisfies the differential equation if

$$E\left(-p^2 \cos pt - 2kp \sin pt + \mu \cos pt\right)$$
$$+ F\left(-p^2 \sin pt + 2kp \cos pt + \mu \sin pt\right) = f \cos pt.$$

Hence we must choose E and F so as to make the last equation an identity for all values of t, i.e. equate to zero the coefficients of $\cos pt$ and $\sin pt$. Therefore

$$(\mu - p^2) E + 2kpF - f = 0,$$

and

$$-2kpE + (\mu - p^2) F = 0 ;$$

whence

$$\frac{E}{\mu - p^2} = \frac{F}{2kp} = \frac{f}{(\mu - p^2)^2 + 4k^2 p^2}.$$

Therefore the required integral of the given differential equation is

$$x = f \frac{(\mu - p^2) \cos pt + 2kp \sin pt}{(\mu - p^2)^2 + 4k^2 p^2},$$

or

$$x = f \frac{\cos (pt + \alpha)}{\{(\mu - p^2)^2 + 4k^2 p^2\}^{\frac{1}{2}}},$$

where

$$\tan \alpha = \frac{2kp}{p^2 - \mu}.$$

Hence, in the case in which $\mu > k^2$, the complete solution is, from **7·8**,

$$x = Ce^{-kt} \cos (nt + \epsilon) + \frac{f \cos (pt + \alpha)}{\{(\mu - p^2)^2 + 4k^2 p^2\}^{\frac{1}{2}}},$$

where $n^2 = \mu - k^2$.

The two terms represent the free and the forced oscillations, and owing to the damping the free oscillations die away and the forced oscillations alone persist.

EXAMPLES

1. A particle of mass m is moving in the axis of x under a central force $\mu m x$ to the origin. When $t=2$ seconds, it passes through the origin, and when $t=4$ seconds, its velocity is 4 feet per second.

Determine the motion and shew that, if the complete period is 16 seconds, the semi-amplitude of the path is $\dfrac{32\sqrt{2}}{\pi}$ feet.   [S. 1926]

2. A mass of 1 lb. is hung on to a light spiral spring and produces a static deflection of $1\frac{1}{2}$ inches. A mass of 1 lb. is suddenly added to the original mass. (i) Find the maximum elongation produced; (ii) shew that the time of an oscillation of the whole mass is approximately $\frac{5}{9}$ sec.   [S. 1911]

3. A body is suspended from a fixed point by a light elastic string of natural length l whose modulus of elasticity is equal to the weight of the body and makes vertical oscillations of amplitude a. Shew that, if as the body rises through its equilibrium position it picks up another body of equal weight, the amplitude of the oscillation becomes $(l^2 + \frac{1}{2}a^2)^{\frac{1}{2}}$.   [S. 1921]

4. A mass m hangs from a fixed point by means of a light spring, which obeys Hooke's law, the mass being given a small vertical displacement. If n is the number of oscillations per second in the ensuing simple harmonic motion, and if l is the length of the spring when the system is in equilibrium, find the natural length of the spring, and shew that, when the spring is extended to double its natural length, the tension is $m\,(4\pi^2 n^2 l - g)$.   [S. 1925]

5. A heavy particle of mass m is attached to one end of an elastic string of natural length a, whose other end is fixed at O. The particle is let fall from rest at O. Shew that part of the motion is simple harmonic, and that, if the greatest depth of the particle below O is $a \cot^2 \frac{1}{2}\theta$, the modulus of elasticity of the string is $\frac{1}{2}mg \tan^2 \theta$, and that the particle attains this depth in time

$$\sqrt{\frac{2a}{g}}\,\{1 + (\pi - \theta) \cot \theta\},$$

where θ is a positive acute angle.   [M. T. 1915]

6. A horizontal board is made to perform simple harmonic oscillations horizontally, moving to and fro through a distance 30 inches and making 15 complete oscillations per minute. Find the least value of the coefficient of friction in order that a heavy body placed on the board may not slip.   [S. 1918]

7. An elastic string is stretched between two fixed points A and B in the same vertical line, B being below A. Prove that, if a particle is fixed to a point P of the string and released from rest in that position, it will oscillate with simple harmonic motion of period $t\sqrt{\mu}$ and of amplitude μa, where t is the period and a the amplitude when P coincides with the mid-point of AB, and $\mu = 4AP \cdot PB/AB^2$. The string may be assumed taut throughout. [S. 1924]

8. A string AB consists of two portions AC, CB of unequal lengths and elasticities. The composite string is stretched and held in a vertical position with the ends A and B secured. A particle is attached to C, and the steady displacement of C is found to be δ. Shew that a further small vertical displacement of C will cause the particle to execute a Simple Harmonic Motion, and that the length of the equivalent simple pendulum is δ. Both portions of the string are assumed to be in tension throughout, and the weight of the string may be neglected. [M. T. 1920]

9. A light elastic string is stretched between two points in the same vertical line, distant l apart. The tension in the string is F. A body, whose weight is small compared with F, is attached to the mid-point of the string, causing it to sink a distance d. Shew that the periodic time, T_1, of small vertical oscillations of the body is the same as that of a simple pendulum of length d.

If the periodic time of small horizontal oscillations of the body is T_2, shew that the mass of the body is approximately

$$4\,\frac{Fd}{gl}\left(\frac{T_2}{T_1}\right)^2.$$ [M. T. 1925]

10. On a given day the depth at high water over a harbour bar is 32 ft., and at low water $6\frac{1}{4}$ hours earlier it is 21 ft. If high water is due at 3.20 p.m., what is the earliest time at which a ship drawing 28 ft. 6 ins. can cross the bar, assuming the rise and fall of the tide to be simple harmonic? [S. 1917]

11. A particle is constrained to move along a straight line, and is attracted towards a fixed point O in that line by a force proportional to its distance from O. It is subjected in addition to a constant force X, acting in the same straight line away from O, and of magnitude sufficient to hold the particle in equilibrium at another point A. Shew that the most general motion possible to the particle is a simple harmonic oscillation, of arbitrary magnitude and phase, about the point A as centre.

The particle being initially at rest at O, the force X is applied and maintained constant for an interval t, after which it ceases. For what value of t (expressed as a fraction of the natural period T) will the particle arrive at A with zero velocity? [M. T. 1927]

12. A ship is rolling with a period of 10 secs. A man at the masthead 100 feet above the deck is swung to and fro 25 feet on either side of the vertical with a motion which is approximately horizontal and simple harmonic. The man weighs 200 lb. and his horizontal hold failing at 50 lb. he is thrown off the mast. The width of the deck being 80 feet, prove that he falls clear of the ship.

[Assume $\pi^2 = 10$ and $g = 32$ f.s. units.]   [M. T. 1913]

13. A ship is making n complete rolls a minute and the motion of the masthead h feet above sea level may be taken as a horizontal simple harmonic motion of total extent $2a$. When at a distance x from the mean position a weight falls from the masthead. Find where it will hit the water, and prove that the distance of this point from the ship will be a maximum when

$$x = a\left(1 + \frac{hn^2}{1440}\right)^{-\frac{1}{2}}, \text{ approx.}$$ [S. 1927]

14. A particle when hanging in equilibrium at the end of a light elastic string stretches it a distance a. Prove that the period of vibration of the particle in a vertical line through its equilibrium position is the same as that of a simple pendulum of length a.

A light endless elastic string of unstretched length $2b$ passes over two small smooth pegs on the same level distant b apart. A particle is attached to a point on the string and when the particle is in equilibrium the string forms the three sides of an equilateral triangle. Prove that the period of vibration of the particle in a vertical line is the same as that of a pendulum of length $\frac{2\sqrt{3}}{7}b$.   [S. 1915]

15. A mass of 12 lb. hangs from a long elastic string which extends 0·25 inch for every pound of load. The string and the given mass are moving upwards in relative equilibrium with uniform velocity 2 feet per second, when the upper end of the string is suddenly brought to rest. Find the distance through which the mass will oscillate.   [S. 1918]

16. According to Hesiod the anvil of Vulcan would take 9 days and 9 nights to fall from the Earth to the realms of Hades. Placing Hades at the centre of the Earth and assuming that the acceleration downwards varies directly as the distance from the centre (and is 32 ft./sec.^{2} at the Earth's surface), shew that Hesiod's figures would give a value of about 15×10^8 miles for the Earth's radius.   [S. 1924]

17. Prove that if a point move in an arc of a parabola having the vertex as middle point so that the motion of the projection of the point on the axis of the parabola is simple harmonic, then the motion of the projection of the point on the directrix is also simple harmonic and of double the period.   [S. 1924]

18. A heavy particle hangs at one end of a light elastic string which is such that the period of a small vertical oscillation of the particle is $2\pi T$. The string is moving vertically upwards with uniform velocity gT_0 and the particle is in relative equilibrium. Shew that, if the upper end of the string is suddenly fixed, the string will become slack if T_0 is greater than T, and that in this case the new motion has a period

$$2\left(\pi - \cos^{-1} T/T_0\right) T + 2\left(T_0{}^2 - T'^2\right)^{\frac{1}{2}}. \qquad \text{[S. 1926]}$$

19. A ring of mass m can slide on a smooth circular wire of radius a in a horizontal plane. The ring is fastened by an elastic string to a point in the plane of the circle at a distance $c\,(>a)$ from its centre. Shew that if the ring makes small oscillations about its position of equilibrium the period is $2\pi \left\{\dfrac{mla\,(c-a)}{\lambda c\,(c-a-l)}\right\}^{\frac{1}{2}}$, where λ is the modulus of elasticity of the string and $l < (c-a)$ is its natural length. [S. 1924]

20. A particle of mass m lies upon a smooth horizontal table and is attached to three points upon the table, at the vertices of an equilateral triangle of side $2a$, by means of three strings of natural lengths l, l' and l' and of moduli λ, λ' and λ' respectively. Shew that if the particle can rest in equilibrium at the centre of the triangle, then

$$2a\,(\lambda/l - \lambda'/l') = (\lambda - \lambda')\sqrt{3}.$$

Find also the period of a small oscillation of the particle in the line of the string of natural length l. [S. 1926]

21. A particle of unit mass is tied by four equal elastic strings of natural length l and modulus of elasticity λ to the corners of a square. If the particle is displaced a small distance towards one of the corners and then set free, prove that the time of a small oscillation is $\pi\sqrt{\dfrac{al}{\lambda\,(a-l)}}$, where a is the length of the diagonal of the square and a is so much greater than l that the strings remain stretched. [S. 1909]

22. It is required to bring to rest a weight W which has fallen freely from a height h by means of the direct pull of a rope of modulus λ, one end of which is attached to it and the other to a point at a variable height vertically above. Find the minimum length of rope if the tension is not to exceed a given value T. Shew that with this length of rope the distance in which the weight is stopped is

$$\frac{2Wh}{T-2W}. \qquad \text{[S. 1921]}$$

23. A particle is hung at the end D of a light string CD knotted at C to two equal light strings AC, CB fastened at points A, B at the same level. Find the equations of motion for small oscillations of the particle

in the vertical plane through AB, and in the vertical plane through C perpendicular to AB, and integrate them.

If a is the depth of C below AB in equilibrium, and $CD = b$, shew that, when $a = 3b$, the particle may be made to describe an arc of a parabola.

[M. T. 1922]

ANSWERS

2.   4·5 ins.   4. $l - g/4\pi^2 n^2$.   6.   ·098.   10.   0 h. 56 m. 57 s.

11. $\frac{1}{6}T$.   15.   4·24 ins.   20. $2\pi \left\{ m \Big/ \left(\dfrac{\lambda}{l} + \dfrac{2\lambda'}{l'} - \dfrac{3\sqrt{3}\,\lambda'}{4a} \right) \right\}^{\frac{1}{2}}$.

22. $2\lambda W h/(T^2 - 2TW)$.

Chapter VIII

MOTION UNDER CONSTRAINT

8·1. A particle may be constrained to move along a given curve or surface, and the constraint may be one-sided, as for example when a heavy particle slides on the inside of a spherical surface and is free to break contact with the surface on the inside of the sphere but cannot get outside. There will then be a normal pressure inwards exerted by the sphere on the particle so long as contact persists, and the pressure will vanish at the point where the particle leaves the surface. On the other hand if the constraint is two-sided as when a particle moves in a fine tube, or a bead moves along a wire, then the normal reaction may vanish and change sign but the particle persists in the prescribed path.

8·2. Motion of a Heavy Particle on a Smooth Curve in a Vertical Plane. The motion is determined by the tangential and normal components of acceleration. The beginner may find it useful in such problems as this to make two diagrams, one showing the components of acceleration multiplied by the mass and the other showing the forces. It is then only necessary to realize that the two diagrams are equivalent representations of the same vector, so that the resolved parts in any assigned direction in the two diagrams are equal.

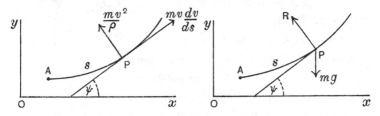

If m is the mass of the particle the forces acting on it are the weight mg and the reaction R along the normal. The components of acceleration are $v\,dv/ds$ along the tangent and v^2/ρ

along the inward normal (**5·11**). Hence, by resolving along the tangent, we get

$$mv\,dv/ds = -\,mg\sin\psi = -\,mg\,dy/ds,$$

therefore, by integration,

$$\tfrac{1}{2}mv^2 = C - mgy\,;$$

or, if v_0 is the velocity when the ordinate is y_0, we have

$$\tfrac{1}{2}m\,(v^2 - v_0{}^2) = mg\,(y_0 - y)\ \dots\dots\dots\dots(1).$$

This is the equation of energy and might have been written down at once; for since the curve is smooth no work is done by the reaction R in any displacement, so the increase in kinetic energy is equal to the work done by the weight.

Again, resolving along the normal, we get

$$mv^2/\rho = R - mg\cos\psi\ \dots\dots\dots\dots\dots(2).$$

Substituting for v from (1), we have

$$R = mg\cos\psi + m\,\{v_0{}^2 + 2g\,(y_0 - y)\}/\rho\ \dots\dots\dots(3).$$

Assuming that the form of the curve is given, the values of ρ and ψ at any point can be determined, and thus R is known; and if we equate to zero the value of R we shall have an equation to determine the point, if any, at which the particle leaves the curve.

8·3. Motion of a Heavy Particle, placed on the outside of a Smooth Circle in a Vertical Plane and allowed to slide down. If the particle starts from Q at an angular distance α from the highest point A, and a is the radius of the circle and v the velocity at P where the angular distance from A is θ, then, from **8·2** (1),

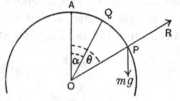

$$v^2 = 2ga\,(\cos\alpha - \cos\theta).$$

Also by resolving along the inward normal

$$mv^2/a = mg\cos\theta - R,$$

where R is the outward reaction of the curve.

Therefore $\qquad R = mg\,(3\cos\theta - 2\cos\alpha),$

shewing that the pressure vanishes, and that the particle flies off the curve, when $\cos\theta = \tfrac{2}{3}\cos\alpha$.

8·31. Motion in a Vertical Plane of a Heavy Particle attached by a Fine String to a Fixed Point. Suppose that the particle starts with velocity u from its lowest position B. If v is the velocity at P and θ is the angle that the string makes with the vertical, the equation of energy is

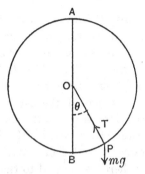

$$\tfrac{1}{2} m\,(v^2 - u^2) = - mg\,a\,(1 - \cos\theta)...(1),$$

and by resolving along the inward normal

$$mv^2/a = T - mg\cos\theta,$$

where T is the tension of the string.

Therefore

$$T = m\,(3g\cos\theta - 2g + u^2/a) \quad ....(2).$$

To find the height of ascent we put $v = 0$ in (1), and get

$$2ga\cos\theta = 2ga - u^2 \quad ....(3),$$

and by putting $T = 0$ in (2), we find that the tension vanishes when

$$3ga\cos\theta = 2ga - u^2 \quad ....(4).$$

We have the following cases:

(i) If $u^2 < 2ga$, the string does not reach the horizontal position and the tension does not vanish.

(ii) If $u^2 = 2ga$, the string just reaches the horizontal position, the tension vanishes for $\theta = \tfrac{1}{2}\pi$, and the particle swings through a quadrant on each side of the vertical.

(iii) If $2ga < u^2 < 5ga$, we find that there is a value of θ, an obtuse angle, given by (4) smaller than that given by (3), so that the string becomes slack before the velocity vanishes and the particle will fall away from the circular path and move in a parabola till the string again becomes taut.

(iv) If $u^2 = 5ga$, the tension just vanishes in the highest position, but v does not vanish, so that circular motion persists.

(v) If $u^2 > 5ga$, neither v nor T vanish.

This is an example of a one-sided constraint; if instead of the problem of a particle attached to a string, we consider that of a bead sliding on a wire, we find that if $u^2 = 4ga$ the bead will reach the highest point of the wire and for any greater value of u it will describe the complete circle.

8·4. Cycloidal Motion. A cycloid is a curve traced out by a point on the circumference of a circle as the circle rolls along a straight line. Let P be the point on the circumference, C the centre of the circle, AGD the line on which it rolls, G the point of contact, i.e. the instantaneous centre of rotation. PG

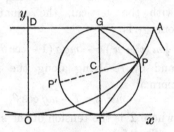

is therefore normal to the path of P and the tangent PT passes through the other end of the diameter through G. Let A be the position of P when it is on the given line, then since the circle rolls we have $AG = \text{arc } GP$, and if PCP' is a diameter and $GD = \text{arc } GP'$, then P' will coincide with D when the circle has turned through an angle π, and P will then be at O, which is called the vertex of the curve, the line AD being called the base. As the rolling proceeds the curve is repeated with a cusp at A and wherever P reaches the fixed line.

Take axes Ox parallel to DA and Oy along OD, and let x, y be the coordinates of P and a the radius of the circle. Let the angle $PTx = \psi$. Then the angle PGT in the alternate segment is also ψ and PCT is 2ψ, so that

$$x = OT + CP \sin 2\psi = a(2\psi + \sin 2\psi),$$

and $$y = CT - CP \cos 2\psi = a(1 - \cos 2\psi).$$

These are the 'parametric' equations of the cycloid. For the intrinsic equation, if s denote the arc OP, we have

$$ds^2 = dx^2 + dy^2$$
$$= 4a^2\{(1 + \cos 2\psi)^2 + \sin^2 2\psi\}\,d\psi^2$$
$$= 8a^2(1 + \cos 2\psi)\,d\psi^2 = 16a^2\cos^2\psi\,d\psi^2,$$

so that $$ds = 4a \cos\psi\,d\psi,$$

and by integration $$s = 4a \sin\psi \quad\dots\dots\dots\dots\dots\dots(1);$$

and this is the intrinsic equation, no constant of integration being required since s and ψ vanish together.

We notice that the radius of curvature

$$\rho = ds/d\psi = 4a \cos \psi = 2PG,$$

so that if PG be produced to Q so that $GQ = PG$, then Q is the centre of curvature.

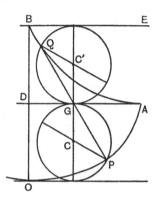

Also if we draw an equal circle to pass through Q and touch AG at G, it is easily seen that the locus of centres of curvature or evolute of the cycloid could be constructed by rolling this second circle along a parallel line EB. A cusp of one cycloid corresponding to a vertex of the other and vice versa.

Now let us consider the motion of a particle under gravity on a smooth cycloid in a vertical plane with its base horizontal and vertex downwards.

Resolving along the tangent we have $m\ddot{s} = -mg \sin \psi$; but

$$s = 4a \sin \psi,$$

therefore $\ddot{s} = -gs/4a$.

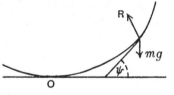

This equation represents a periodic motion, the time of oscillation being $2\pi \sqrt{(4a/g)}$. It follows that the particle oscillates on either side of the vertex with a period that is independent of the amplitude. This property is called the 'isochronism of the cycloid.'

We recall that the formula $2\pi \sqrt{(l/g)}$ obtained for the period of vibration of a simple pendulum depended on the amplitude being so small that θ can be used for $\sin \theta$, and we now see that if the 'bob' of the pendulum could be made to move in a cycloidal path the formula for the period of vibration would be independent of the amplitude. This can be attained by making the string that supports the bob wrap and unwrap itself on cycloidal arcs.

Thus in the second figure, from (1)

$$\text{the arc } AQ = 4a \sin QGD$$
$$= QP,$$

while $\qquad\qquad$ arc $AB = 4a$.

Hence if a string of length $4a$ had one end fastened at B and were wrapt round the curves BA and BA' alternately, the

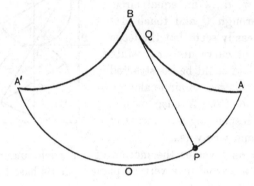

other end P would trace out the cycloid AOA', and a particle attached to P moving under gravity and the tension of the string would oscillate in the time $2\pi \sqrt{(4a/g)}$ whatever be the amplitude of the oscillation within the limits AOA'.

EXAMPLES

1. A heavy particle of weight W, attached to a fixed point by a light inextensible string, describes a circle in a vertical plane. The tension in the string has the values mW and nW, respectively, when the particle is at the highest and lowest points in its path.

Shew that $n = m + 6$. $\qquad\qquad\qquad\qquad\qquad\qquad\qquad\qquad$ [M. T. 1927]

2. A heavy particle slides under gravity down the inside of a smooth circular tube held in a vertical plane. The particle starts at the highest point with the velocity it would acquire if it fell down the radius, prove that when in the subsequent motion the vertical component of its acceleration is a maximum the pressure on the curve is equal to twice the weight of the particle. $\qquad\qquad\qquad\qquad\qquad\qquad\qquad\qquad\qquad$ [S. 1919]

3. A heavy particle, hanging from a fixed point by a light inextensible string of length a, is projected horizontally with velocity V. Shew that during the circular motion the tension of the string at any time is proportional to the depth of the particle at that moment below a certain horizontal line ; and find the values between which V must lie that the string may become slack.　　　　　　　　　　　　　　[M. T. 1916]

4. A particle is projected along the inner side of a smooth circle of radius a, the velocity at the lowest point being u. Shew that if $u^2 < 5ga$ the particle will leave the circle before arriving at the highest point and will describe a parabola whose latus rectum is $2\,(u^2-2ga)^3/27g^3a^2$.

[S. 1916]

5. A particle slides down the surface of a smooth fixed sphere of radius a, being slightly displaced from rest at the highest point. Find where it will leave the sphere, and shew that it will afterwards describe a parabola of latus rectum $\frac{16}{27}a$, and that it will strike the horizontal plane through the lowest point of the sphere at a distance $\dfrac{5\,(\sqrt{5}+4\,\sqrt{2})\,a}{27}$ from the vertical diameter.　　　　　　　　　　　　　　[S. 1918]

6. Two beads connected by a string are held at rest on a vertical circular wire with the string horizontal, and above the centre. Their masses are m, m', and the string subtends an angle $2a$ at the centre. If the beads are released, shew that the tension of the string when it makes an angle θ with the horizontal is

$$\frac{2mm'g \tan a \cos \theta}{m+m'}.$$

[S. 1917]

7. Two equal particles are tied together by a light string of length $\pi a/2$ and rest in equilibrium on the surface of a smooth circular cylinder in a plane perpendicular to the axis of the cylinder, which is horizontal, the radius of the cylinder being a. The particles are then slightly displaced in a plane perpendicular to the axis of the cylinder. Prove that when the lower of the two particles leaves the surface, the pressure on the other is rather more than three-fourths of its weight.　　　　　　　　　[S. 1912]

8. A particle starts from rest at any point P in the arc of a smooth cycloid whose axis is vertical and vertex A downwards; prove that the time of descent to the vertex is $\pi\,\sqrt{\dfrac{a}{g}}$, where a is the radius of the generating circle.

Shew also that if the particle is projected from P downwards along the curve with velocity equal to that with which it reaches A when starting from rest at P, it will now reach A in half the time taken in the preceding case.　　　　　　　　　　　　　　　　　　　　　　　　[S. 1915]

9. A particle is constrained to move under gravity on a parabola with axis vertical and vertex upwards. Shew that the pressure on the curve is numerically $m(u_0^2 - gl)/\rho$, where u_0 is the velocity at the vertex, $2l$ the latus rectum and ρ the radius of curvature at any point.

What is the meaning of the special case $u_0^2 = gl$?   [M. T. 1927]

10. A heavy particle P slides on a smooth curve of any form in a vertical plane. The centre of curvature at P is Q, and R is on the same vertical as Q and at the level of zero velocity. Shew that the acceleration makes with the normal an angle $\tan^{-1}(\frac{1}{2}\tan PRQ)$.   [S. 1912]

11. A smooth wire is bent into the form $y = \sin x$ and placed in a vertical plane with the axis of x horizontal. A bead of mass m slides down the wire starting from rest at $x = \frac{1}{4}\pi$. Shew that the pressure on the wire as the bead passes through the origin is $mg/\sqrt{2}$, and find the pressure as it passes through $x = -\frac{1}{2}\pi$.   [S. 1921]

12. A bead moves on a smooth circular wire under the action of forces tending to the corners of a regular polygon concentric with the circle. The forces vary as the distance and are equal at equal distances. Prove that the pressure on the wire is constant.   [S. 1901]

13. A particle of mass m is attached by a string to a point on the circumference of a fixed circular cylinder of radius a whose axis is vertical, the string being initially horizontal and tangential to the cylinder. The particle is projected with velocity v at right angles to the string along a smooth horizontal plane so that the string winds itself round the cylinder.

Shew (i) that the velocity of the particle is constant,

(ii) that the tension in the string is inversely proportional to the length which remains straight at any moment,

(iii) that if the initial length of the string is l and the greatest tension the string can bear is T, the string will break when it has turned through an angle

$$l/a - mv^2/aT.$$ [S. 1927]

14. A switchback railway consists of straight stretches smoothly joined by circular arcs, the whole lying in a vertical plane. Shew that a car started on a level stretch and running freely will leave the track if the downward gradient exceed $\cos^{-1}\frac{2}{3}$ at any point; but that if braking is available up to half the weight of the car, gradients of about $77°$ are admissible.

A level and a straight descending stretch are smoothly joined by an arc of radius a. Two equal cars without brakes are joined by a cable of length $2a$. Shew that the greatest admissible gradient such that the second car does not leave the track is $\cos^{-1}\frac{12}{13}$.

[Neglect the size of the cars, and resistance to motion other than braking.]   [S. 1924]

15. A small ring fits loosely on a rough spoke (length a) of a wheel which can turn about a horizontal axle and the ring is originally at rest in contact with the lowest point of the rim : if the wheel is now made to revolve with uniform angular velocity ω, prove that the angle θ through which the wheel will turn before the ring slides is given by the equation

$$\cos(\theta-\lambda)/\cos\lambda + \omega^2 a/g = 0,$$

where λ is the angle of friction.   [S. 1910]

ANSWER

11. $mg\,(3+\sqrt{2})$.

Chapter IX

THE LAW OF REACTION. GENERAL PRINCIPLES

9·1. So far we have been concerned with the motion of a single particle. When two or more particles are moving in such a manner that the motion of any one is affected by the presence of the others we have to make use of another law enunciated by Newton, viz. *Action and Reaction are equal and opposite*, or, *the actions of two bodies on one another are always equal and opposite*. In explicit terms this means that if a body A exerts a force F on a body B, then B exerts an equal force F on A but in the opposite direction. Consequently the momentum communicated to A by the action of B is equal and opposite to the momentum communicated to B by the action of A.

Consider the case of a system of bodies, attracting or repelling each other or acting on one another by contact, or through connections by means of strings or rods, either for a finite time or by instantaneous impulses. In this case any momentum which is produced or destroyed in any assigned direction is accompanied by the production or destruction of an equal momentum in the opposite direction.

Hence it follows that, if no external forces act on a system of bodies, the total momentum of the system in any assigned direction remains constant.

This is **the principle of conservation of linear momentum.**

9·2. Motion of a system of particles. Let (x_1, y_1), (x_2, y_2), etc. be the coordinates of a system of particles of masses m_1, m_2, etc. Let the particles be subject to given external forces whose components parallel to the axes are

$$X_1, Y_1, \quad X_2, Y_2, \text{ etc.}$$

and also to internal actions and reactions due to the mutual actions of the particles upon one another of which the components on m_1, m_2, etc. are

$$X_1', Y_1', \quad X_2', Y_2', \text{ etc.}$$

Writing down the equations of motion for the separate particles, we have

$$\left.\begin{array}{ll} m_1\ddot{x}_1 = X_1 + X_1', & m_1\ddot{y}_1 = Y_1 + Y_1' \\ m_2\ddot{x}_2 = X_2 + X_2', & m_2\ddot{y}_2 = Y_2 + Y_2' \\ \text{etc.} & \text{etc.} \end{array}\right\} \quad\dots\dots\dots(1).$$

Whence by addition

$$\left.\begin{array}{l} \Sigma m\ddot{x} = \Sigma X + \Sigma X' \\ \Sigma m\ddot{y} = \Sigma Y + \Sigma Y' \end{array}\right\} \quad\dots\dots\dots\dots(2).$$

and

Now by the law of reaction the internal actions and reactions are equal and opposite in pairs, so that the sums of their resolved parts in any direction must vanish, i.e. $\Sigma X' = 0$ and $\Sigma Y' = 0$.

Hence equations (2) reduce to

$$\Sigma m\ddot{x} = \Sigma X, \quad \Sigma m\ddot{y} = \Sigma Y \dots\dots\dots\dots(3),$$

or, in words, *the rate of change of the linear momentum of the whole system in any prescribed direction is equal to the sum of the resolved parts of the external forces in that direction.*

It follows that if there is a direction, say the axis of x, in which the sum of the resolved parts of the external forces is zero, i.e. $\Sigma X = 0$; then, by integrating $\Sigma m\ddot{x} = 0$ we get

$$\Sigma m\dot{x} = \text{const.,}$$

i.e. the linear momentum in that direction is constant. This is again **the principle of conservation of linear momentum.**

Again, from equations (1), by multiplying each y equation by the corresponding x and each x equation by the corresponding y and subtracting, we deduce that

$$m_1(x_1\ddot{y}_1 - y_1\ddot{x}_1) = x_1Y_1 - y_1X_1 + x_1Y_1' - y_1X_1',$$
$$m_2(x_2\ddot{y}_2 - y_2\ddot{x}_2) = x_2Y_2 - y_2X_2 + x_2Y_2' - y_2X_2',$$
$$\text{etc., etc.,}$$

whence by addition

$$\Sigma m(x\ddot{y} - y\ddot{x})$$
$$= \Sigma(xY - yX) + \Sigma(xY' - yX').$$

Now the moment about the origin of a vector whose components are X, Y located at the point (x, y) is $xY - yX$; hence $\Sigma(xY' - yX')$ is the sum of the

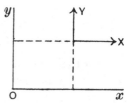

moments about the origin of the internal actions and reactions, which are equal and opposite in pairs. Therefore

$$\Sigma\,(xY' - yX') = 0,$$

and we have

$$\Sigma m\,(x\ddot{y} - y\ddot{x}) = \Sigma\,(xY - yX) \quad\ldots\ldots\ldots\ldots(4).$$

This may also be written

$$\frac{d}{dt}\,\Sigma m\,(x\dot{y} - y\dot{x}) = \Sigma\,(xY - yX),$$

or, in words, *the rate of change of moment of momentum of the system about any fixed origin (or axis) is equal to the sum of the moments of the external forces about that origin (or axis).*

If the sum of the moments of the external forces about any fixed axis is zero, it follows that the moment of momentum of the system about that axis is constant. This is **the principle of conservation of moment of momentum.**

Moment of momentum is frequently called **angular momentum.**

9·21. Effective Forces. The product of the mass and the acceleration of a particle is called the *effective force* of the particle, and the principles embodied in equations (3) and (4) of the last article are that for any system of particles the effective forces are the exact equivalent of the external forces acting on the system.

9·3. Motion of the Centre of Gravity. Independence of Translation and Rotation. If M be the whole mass of the system of particles and $\bar{x}$, $\bar{y}$ the coordinates of the centre of gravity G, the usual formulae for the position of the centre of gravity are

$$M\bar{x} = \Sigma mx, \quad M\bar{y} = \Sigma my\ldots\ldots\ldots\ldots\ldots(1).$$

By differentiating these equations we get

$$M\dot{\bar{x}} = \Sigma m\dot{x}, \quad M\dot{\bar{y}} = \Sigma m\dot{y},$$

shewing that the linear momentum of the system is the same as that of a particle whose mass is the whole mass, moving with the velocity of the centre of gravity.

A second differentiation gives
$$M\ddot{\bar{x}} = \Sigma m\ddot{x}, \quad M\ddot{\bar{y}} = \Sigma m\ddot{y},$$
therefore from (3) of **9·2**
$$M\ddot{\bar{x}} = \Sigma X, \quad M\ddot{\bar{y}} = \Sigma Y, \dots\dots\dots\dots(2),$$
and these equations shew that *the motion of the centre of gravity G is the same as if all the mass were collected into a particle at G and all the external forces were moved parallel to themselves to act at G.*

It follows that if the system is not acted upon by external forces its centre of gravity is either at rest or moving with constant velocity, for in this case the integration of equation (2) gives $M\dot{\bar{x}} = $ const., $M\dot{\bar{y}} = $ const.

Again in equation (4) of **9·2** we put
$$x = \bar{x} + x', \quad y = \bar{y} + y',$$
so that x', y' denote coordinates of the particle of mass m relative to axes through the centre of gravity G; then, from (1),
$$\Sigma m x' = \Sigma m y' = 0,$$
so that by differentiation $\Sigma m\ddot{x}' = \Sigma m\ddot{y}' = 0$, and (4) of **9·2** becomes
$$\Sigma m\{(\bar{x} + x')(\ddot{\bar{y}} + \ddot{y}') - (\bar{y} + y')(\ddot{\bar{x}} + \ddot{x}')\}$$
$$= \Sigma\{(\bar{x} + x')\,Y - (\bar{y} + y')\,X\}.$$

Multiplying out and observing that such terms as
$$\Sigma m\bar{x}\ddot{y}' = \bar{x}\Sigma m\ddot{y}' = 0,$$
there remains
$$M(\bar{x}\ddot{\bar{y}} - \bar{y}\ddot{\bar{x}}) + \Sigma m(x'\ddot{y}' - y'\ddot{x}') = \bar{x}\Sigma Y - \bar{y}\Sigma X + \Sigma(x'Y - y'X).$$

In virtue of (2) this equation reduces to
$$\Sigma m(x'\ddot{y}' - y'\ddot{x}') = \Sigma(x'Y - y'X) \dots\dots\dots\dots(3),$$
shewing that, *in the motion of the system relative to the centre of gravity, the rate of change of moment of momentum about the centre of gravity is equal to the sum of the moments of the external forces about the centre of gravity.*

The results of this article shew that the motion of the centre of gravity and of the system relative to the centre of gravity are independent of one another and this constitutes **the principle of the independence of translation and rotation.**

9·4. Conservation of Energy. Reverting to equations (1) of **9·2**, by multiplying each equation by the corresponding velocity component and adding, we get

$$\Sigma m \, (\dot{x}\ddot{x} + \dot{y}\ddot{y}) = \Sigma \left\{ (X + X') \, \dot{x} + (Y + Y') \dot{y} \right\},$$

or $$\frac{d}{dt} \left\{ \tfrac{1}{2} \Sigma m \, (\dot{x}^2 + \dot{y}^2) \right\} = \Sigma \left\{ (X + X') \, \dot{x} + (Y + Y') \, \dot{y} \right\}.$$

The left-hand side is the rate of increase of the kinetic energy of the system, and the right-hand side is the rate at which all the forces external and internal are doing work. From this we conclude that the increase of kinetic energy in any time is equal to the whole work done. In all cases in which the potential energy depends on the configuration of the system and in which the change of potential energy due to a change of configuration is independent of the manner in which that change is made, the work done will be equal to the loss of potential energy. Whence it follows that

kinetic energy + potential energy = constant.

Kinetic energy may be created by explosions, in which case the visible kinetic energy together with the heat developed are the equivalent of the chemical energy stored in the explosive substance. Similarly kinetic energy may be dissipated by friction when it is converted into heat.

9·5. Kinetic Energy in reference to Centre of Gravity. *The kinetic energy of a system of particles is equal to the kinetic energy of the whole mass moving with the velocity of the centre of gravity together with the kinetic energy of the particles in their motion relative to the centre of gravity.*

Using the notation of **9·3**, we have that the kinetic energy

$$= \tfrac{1}{2} \Sigma m \, (\dot{x}^2 + \dot{y}^2)$$
$$= \tfrac{1}{2} \Sigma m \, \{ (\dot{\bar{x}} + \dot{x}')^2 + (\dot{\bar{y}} + \dot{y}')^2 \}$$
$$= \tfrac{1}{2} M \, (\dot{\bar{x}}^2 + \dot{\bar{y}}^2) + \dot{\bar{x}} \Sigma m \dot{x}' + \dot{\bar{y}} \Sigma m \dot{y}' + \tfrac{1}{2} \Sigma m \, (\dot{x}'^2 + \dot{y}'^2).$$

But $\Sigma m x' = \Sigma m y' = 0$, so that $\Sigma m \dot{x}' = \Sigma m \dot{y}' = 0$, and the kinetic energy $= \tfrac{1}{2} M \, (\dot{\bar{x}}^2 + \dot{\bar{y}}^2) + \tfrac{1}{2} \Sigma m \, (\dot{x}'^2 + \dot{y}'^2)$, which proves the proposition.

9·6. In the last few articles we have distinguished between external and internal forces. Whether any particular force is to be classed as external or internal depends on the way in which we regard the particles composing the system. Thus if we are considering the motion of a single particle A, then all the forces acting upon it are classed as 'external forces,' but if we are considering a system composed of two particles A and B then the forces exerted by B on A and A on B are internal forces. The force of gravity is an external force in considering the motion of a body relative to the earth, but in considering the motion of the earth and moon regarded as one system moving about the sun the gravitational pull of the earth on the moon or the moon on the earth is an internal force.

9·7. Rigid Bodies. A rigid body is considered to be an aggregation of particles bound together by forces of cohesion and internal mutual attractions which are in all cases equal and opposite. The results obtained in this chapter for a system of particles are therefore true for a rigid body. Consequently the motion of the centre of gravity of such a body does not depend upon its size or shape but only upon its mass and the resultant of the external forces acting upon it. It is only when we are concerned with rotational motion relative to the centre of gravity that the size and shape come into consideration, for upon them depend the moment of momentum and the moments of the forces.

9·8. In the chapter on rectilinear motion when we applied the second law of motion to rigid bodies or cars and trains we considered every particle of the body to be moving with the same acceleration, and any assumptions there made which might seem to need further justification find it in the principles established in **9·3** of this chapter. Moreover many problems about rigid bodies not involving rotational properties may be solved by elementary methods in virtue of the principle that *the motion of the centre of gravity G is the same as if all the mass were collected into a particle at G and all the external forces were moved parallel to themselves to act at G.*

As an example let us *find the condition that a car may upset when rounding a curve.*

Let v be the velocity and m the mass. We are concerned with the forces in planes at right angles to the direction of motion.

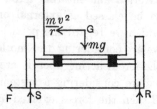

If r is the radius of curvature of the curve described by the centre of gravity G, then G has an acceleration v^2/r inwards along the normal to its path. Therefore the resultant of all the forces perpendicular to the direction of motion must be mv^2/r. But the only force in this direction is friction on the wheels. So if F is the total friction across the track we have $F = mv^2/r$. Let R, S be the upward pressure of the ground on outer and inner wheels, h the height of G and $2a$ the distance between the wheels.

Then by resolving vertically, $R + S = mg$, and by taking moments about G (since there is no moment of momentum about G) we get

$$(R - S)\, a = Fh,$$

therefore

$$R - S = mv^2h/ra.$$

Hence

$$2R = mg + mv^2h/ra,$$

and

$$2S = mg - mv^2h/ra.$$

We conclude that if $v^2 > gra/h$ the car will overturn outwards, since this condition implies a negative pressure of the inner wheels on the ground which is an impossibility.

9·81. The necessity for sideways friction F can be obviated by banking up the track. If the track be banked up to make an angle θ with the horizontal and there is no sideways friction (i.e. no tendency to side-slip), we get, by resolving horizontally and vertically,

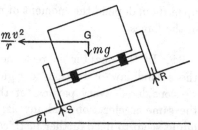

$$mv^2/r = (R + S) \sin \theta,$$

and

$$mg = (R + S) \cos \theta.$$

Therefore $v^2 = rg \tan \theta$ gives the velocity at which tendency to side-slip is eliminated for a track of slope θ.

EXAMPLES

1. The wheel axles of a motor car are 4 feet long and the height of the c.g. is 2 feet. Find the speed of the car if in going round a level track of 400 feet radius the inner wheels just leave the ground.   [S. 1916]

2. A skater describes a circle of 40 feet radius with a velocity of 15 feet per second. At what angle must he lean inwards?   [S. 1914]

3. Prove that, in order to allow properly for a curve on a railway line of radius 1320 feet for a train moving at 45 miles an hour, the outer rail must be raised above the inner rail by 5·8 inches. (The rails are 4 feet 8½ inches apart.)   [M. T. 1921]

4. A railway of gauge 5 feet is taken round a curve of ½ mile radius. What 'superelevation' must be given to the outer rail in order that a train travelling round this curve at 30 and at 60 m.p.h. may impose the same side pressure on the inner and outer rail respectively?   [M. T. 1928]

5. A car takes a banked corner of a racing track at a speed V, the lateral gradient α being designed to reduce the tendency to side-slip to zero for a lower speed U. Shew that the coefficient of friction necessary to prevent side-slip for the greater speed V must be at least

$$(V^2 - U^2)\sin\alpha\cos\alpha/(V^2\sin^2\alpha + U^2\cos^2\alpha).   [S. 1925]$$

6. A pile-driver weighing 200 lb. falls through 5 feet and drives a pile which weighs 600 lb. through a distance of 3 inches. Find the average resistance offered to the motion of the pile, assuming that the two remain in contact after the blow.

How many foot-pounds of energy are dissipated during the blow?
          [S. 1926]

7. A shell of mass $(m_1 + m_2)$ is fired with a velocity whose horizontal and vertical components are U, V, and at the highest point in its path the shell explodes into two fragments m_1, m_2. The explosion produces an additional kinetic energy E, and the fragments separate in a horizontal direction: shew that they strike the ground at a distance apart which is equal to

$$\frac{V}{g}\left\{2E\left(\frac{1}{m_1} + \frac{1}{m_2}\right)\right\}^{\frac{1}{2}}.   [S. 1924]$$

8. A gun of mass M fires a shell of mass m horizontally, and the energy of the explosion is such as would be sufficient to project the shell vertically to a height h. Shew that the velocity of recoil of the gun is

$$\{2m^2gh/M(M+m)\}^{\frac{1}{2}}.   [S. 1927]$$

9. A set of n trucks with s feet clear between them are inelastic and are set in motion by starting the end one with velocity V towards the next. Find how long it takes for the last truck to start and the value of the final velocity.   [S. 1927]

10. A battleship of symmetrical form and mass 30,000 tons is moving at 10 miles per hour and fires a salvo of all its eight guns in a direction perpendicular to its motion. If the shells weigh 15 cwt. each, have a muzzle velocity of 2000 feet per second, and are fired at an elevation of 30°, shew that the direction of motion immediately after firing makes an angle of about 1° 21' with that before.   [S. 1922]

11. A ball is dropped from the top of a tower 100 feet high. At the same moment a ball of equal mass is thrown from a point on the ground 50 feet from the foot of the tower so as to strike the first ball when just half-way down. Find the initial velocity of projection of the second ball and the direction of projection. If the two balls coalesce how long will they take to reach the ground?   [S. 1926]

12. A shell of mass 1120 lb., and velocity 1350 feet per second, is fired into a railway truck (containing sand) of mass 20 tons, the direction of motion being parallel to the rails. If the shell fails to penetrate the sand, find the velocity given to the truck and account for the conservation of energy in the phenomenon, specifying how much remains kinetic. How far will the truck run against a constant retarding force of 30 lb. weight per ton?   [M. T. 1921]

13. A pile-driver weighing 2 cwt. falls through 5 feet and drives a pile weighing 6 cwt. through a distance of 4 inches. Find the average resistance to the pile in cwt., assuming the two to remain in contact. Find in foot-pounds the energy dissipated in one stroke.   [S. 1917]

ANSWERS

1. 113·13 f.s.   2. 9° 58' to vertical.   4. 3·45 ins.   6. 1000 lb. wt.; 750.   9. $\frac{1}{2}n(n-1)s/V$; V/n.   11. $20\sqrt{10}$ f.s.; $\tan^{-1}2$ above horizontal; $5(\sqrt{5}-1)/4\sqrt{2}$.   12. $32\frac{38}{41}$ f.s.; 777,896 ft.-lb.; 1265 ft.   13. 7·5 cwt.; 280 ft.-lb.

Chapter X

GENERAL PROBLEMS

10·1. We shall now apply the principles already established to a few simple problems.

The figure shews an arrangement of pulleys, two fixed and two movable. The latter are of masses m_1, m_2 and the strings passing round them can slip without friction, so that the tension of each string is constant throughout its length. Particles of mass m_3, m_4, m_5 are attached to the free ends of the strings. Let the tensions be denoted by T, T'. It is not necessary to use five unknown quantities to denote the accelerations of the five masses, for if f, f', f'' denote the accelerations with which the strings are slipping over the pulleys as indicated in the figure, then the upward accelerations of m_1, m_2, m_3, m_4, m_5 are

$$f', \tfrac{1}{2}(f''-f'), \ f'+f, \ f'-f, \ -f''.$$

Therefore
$$m_1 f' = T' - 2T - m_1 g,$$
$$\tfrac{1}{2}m_2(f''-f') = 2T' - m_2 g,$$
$$m_3(f'+f) = T - m_3 g,$$
$$m_4(f'-f) = T - m_4 g,$$
$$m_5 f'' = m_5 g - T.$$

These equations are sufficient to determine the five unknown quantities f, f', f'', T, T'.

10·2. *A fine smooth wire of mass M forms an equilateral triangle ABC. The triangle can move horizontally in a vertical plane, the uppermost side BC passing through smooth fixed rings in a horizontal line. Beads of masses m and m' are free to slide on the wires BA, CA. The system begins to move with the beads at B and C respectively. Prove that the velocity of the wire at any instant, while both the beads are moving on it, is equal to the difference of the speeds of the beads relative to the wire, and that the acceleration of the wire is*

$$\sqrt{3}\,(m' \sim m)\,g/(4M + 3m + 3m').$$
[S. 1927]

At time t from the start let V be the velocity of the wire and v, v' the velocities of m, m' relative to the wire.

Since there is no external horizontal force acting on the system as a whole the horizontal momentum remains zero.

Therefore

$$MV + m\,(V + v \cos 60°)$$
$$+ m'\,(V - v' \cos 60°) = 0\ldots\ldots(1).$$

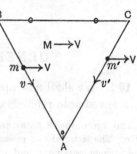

Resolving along BA for m and along CA for m', we have

$$m\,(\dot{v} + \dot{V} \cos 60°) = mg \cos 30°\Big\}$$
$$\text{and}\ \ m'\,(\dot{v}' - \dot{V} \cos 60°) = m'g \cos 30°\Big\}\ \ldots(2).$$

Therefore by subtraction $\dot{v}' - \dot{v} = \dot{V}$.

Integrating this, and noting that the velocities are all zero initially so that there is no constant of integration required, we get $V = v' - v$.

Again differentiating (1) gives

$$(M + m + m')\,\dot{V} = \tfrac{1}{2}m'\dot{v}' - \tfrac{1}{2}m\dot{v},$$

and substituting for $\dot{v}$ and $\dot{v}'$ from (2) we get

$$(M + m + m')\,\dot{V} = \tfrac{1}{4}m'\,(\dot{V} + \sqrt{3}g) + \tfrac{1}{4}m\,(\dot{V} - \sqrt{3}g),$$

so that $$(4M + 3m + 3m')\,\dot{V} = \sqrt{3}\,(m' - m)g.$$

10·3. *A particle of mass m slides down the face of a smooth wedge of inclination α to the horizontal. The wedge is of mass M and it rests on a rough horizontal table. Shew that the pressure on the table is $(M + m \cos^2 \alpha)\,g$ or $(M + m)\,Mg/\{M + m \sin \alpha\,(\sin \alpha - \mu \cos \alpha)\}$, according as the coefficient of friction μ is greater or less than*

$$m \sin \alpha \cos \alpha/(M + m \cos^2 \alpha).$$

Let R be the reaction between the particle and the wedge, S and F the vertical reaction and the friction acting between the table and the wedge.

(i) If the wedge does not move we must have $F < \mu S$.

But, resolving vertically and horizontally for the wedge, we get

$$S = Mg + R \cos \alpha, \quad F = R \sin \alpha\ldots\ldots(1),$$

and by resolving at right angles to the wedge for the particle, we get

$$R = mg \cos \alpha.$$

Therefore $\quad S = (M + m \cos^2 \alpha)\,g \quad$ and $\quad F = mg \sin \alpha \cos \alpha.$

Hence this value of S is the value for the pressure provided that

$$\mu > m \sin \alpha \cos \alpha/(M + m \cos^2 \alpha).$$

(ii) If μ is less than this value there is not enough friction to prevent motion, so motion takes place and the friction $F=\mu S$. Let f be the acceleration of the wedge (from left to right in the figure) and f' the acceleration of the particle relative to the wedge.

Resolving horizontally for the wedge, we get

$$Mf = R \sin \alpha - \mu S \dots\dots\dots\dots\dots\dots\dots(2).$$

Again, the acceleration of the particle m at right angles to the face of the wedge is the same as the acceleration of the wedge in this direction, viz. $f \sin \alpha$; therefore

$$mf \sin \alpha = mg \cos \alpha - R \dots\dots\dots\dots\dots\dots(3).$$

Eliminate f from (2) and (3), and we get

$$R\,(M+m \sin^2 \alpha) = \mu S m \sin \alpha + Mmg \cos \alpha\,;$$

but from (1) $R \cos \alpha = S - Mg,$

therefore $S\,\{M+m \sin \alpha\,(\sin \alpha - \mu \cos \alpha)\} = (M+m)\,Mg,$

which is the required result.

We note that if

$$\mu = m \sin \alpha \cos \alpha / (M + m \cos^2 \alpha),$$

the two values of S are the same.

10·4. Examples of conservation of momentum and energy.

(i) *A bead of mass M can slide on a smooth straight horizontal wire and a particle of mass m is attached to the bead by a light string of length l. The particle is held in contact with the wire with the string taut and is then let fall. Prove that when the string is inclined to the wire at an angle θ the bead will have slipped a distance $ml\,(1-\cos\theta)/(M+m)$ along the wire, and that the angular velocity ω of the string will be given by the equation*

$$(M+m \cos^2 \theta)\, l\omega^2 = 2(M+m)g \sin \theta. \qquad \text{[M. T. 1915]}$$

Suppose that the bead moves a distance x while the string turns through an angle θ. Then the velocity of the bead is $\dot{x}$ and the particle m has a velocity $l\dot{\theta}$ relative to the bead in addition to the velocity of the bead.

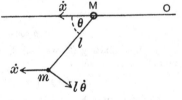

Since there is no external horizontal force, the total horizontal momentum remains zero, therefore

$$M\dot{x} + m\,(\dot{x} - l\dot{\theta} \sin \theta) = 0 \dots\dots\dots\dots\dots\dots(1).$$

Integrating and observing that initially x and θ are both zero, we get

$$Mx + m\,(x + l \cos \theta) = ml,$$

therefore $x = ml\,(1 - \cos \theta)/(M+m).$

This result could also be obtained from the consideration that the centre of gravity of the masses M and m undergoes no horizontal displacement.

Again, by equating the kinetic energy to the work done by gravity we get

$$\tfrac{1}{2}M\dot{x}^2 + \tfrac{1}{2}m\,(\dot{x}^2 + l^2\dot{\theta}^2 - 2\dot{x}\,l\dot{\theta}\sin\theta) = mgl\sin\theta \quad \ldots\ldots\ldots\ldots(2),$$

and eliminating $\dot{x}$ from (1) and (2) gives

$$(M + m\cos^2\theta)\,l\dot{\theta}^2 = 2(M + m)\,g\sin\theta.$$

(ii) *A hemisphere of mass M is free to slide with its base on a smooth horizontal table. A particle of mass m is placed on the hemisphere at an angular distance α from the vertex; to determine the motion, the surface of the hemisphere being smooth.*

Suppose that at time t the particle is at an angular distance θ from the vertex, and that the hemisphere has acquired a velocity v, the velocity of the particle being $a\dot{\theta}$ relative to the hemisphere, where a is the radius.

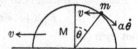

Since there is no external horizontal force acting on the system as a whole the total horizontal momentum remains zero throughout the motion, therefore

$$Mv + m\,(v - a\dot{\theta}\cos\theta) = 0 \ldots\ldots\ldots\ldots\ldots\ldots\ldots\ldots\ldots(1).$$

Also the kinetic energy created is equal to the work done, therefore

$$\tfrac{1}{2}Mv^2 + \tfrac{1}{2}m\,(v^2 + a^2\dot{\theta}^2 - 2va\dot{\theta}\cos\theta) = mga\,(\cos\alpha - \cos\theta) \ \ldots\ldots(2).$$

The result of eliminating v between (1) and (2) is

$$a\dot{\theta}^2 = 2g\,(\cos\alpha - \cos\theta)\,(M + m)/(M + m\sin^2\theta) \ \ldots\ldots\ldots\ldots(3),$$

which gives the velocity of the particle relative to the hemisphere, and then (1) gives the velocity of the hemisphere.

To find at what point the particle leaves the hemisphere, let R denote the mutual reaction between them. Then by resolving horizontally for the hemisphere, we get $M\dot{v} = R\sin\theta.$

But from (1)

$$(M + m)\,\dot{v} = ma\,(\ddot{\theta}\cos\theta - \dot{\theta}^2\sin\theta)$$

and R vanishes when $\dot{v}$ vanishes, i.e. when

$$\ddot{\theta}\cos\theta = \dot{\theta}^2\sin\theta.$$

By differentiating (3) we find $\ddot{\theta}$, then by equating $\ddot{\theta}\cos\theta$ to $\dot{\theta}^2\sin\theta$ we find the equation for θ, viz.

$$m\cos^3\theta - (M + m)\,(3\cos\theta - 2\cos\alpha) = 0,$$

which determines the point at which the particle leaves the surface.

10·5. Examples of Circular Motion. Conical Pendulum.

A particle suspended by a thread from a fixed point is projected so as to describe a horizontal circle with uniform velocity; to prove that the time of revolution is $2\pi\sqrt{(h/g)}$, where h is the depth of the circle below the fixed point.

If ω is the angular velocity, l the length of the string and α its inclination to the vertical, the acceleration of the particle is $\omega^2 l \sin \alpha$ towards the centre of the circle. Therefore if m is the mass of the particle the effective force is $m\omega^2 l \sin \alpha$ towards the centre of the circle and this is the resultant of the weight mg and the tension T. Hence by resolving horizontally and vertically

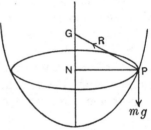

$$m\omega^2 l \sin \alpha = T \sin \alpha,$$
and $$mg = T \cos \alpha.$$
Therefore $$\omega^2 l = g \sec \alpha, \text{ or } \omega^2 = g/h;$$
and the time of revolution $= 2\pi/\omega = 2\pi \sqrt{(h/g)}$.

10·51. The problem of *a particle describing a horizontal circle on the inside of a smooth surface of revolution with a vertical axis* is solved in the same way.

Thus if P be the particle, PN perpendicular to the axis is the radius of the circle described by P. And if the normal to the surface at P meets the axis at G, the forces on the particle in this case are its weight mg and the reaction R of the surface along PG, and as in the last article $\omega^2 . NG = g$.

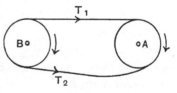

10·6. Transmission of Energy. Power is sometimes transmitted from one rotating shaft or pulley to another by means of a belt passing round both. Thus if a shaft A be made to rotate by an engine, a belt round it and another shaft B will cause the latter to rotate. Let b be the radius and ω the angular velocity of the shaft B, and let T_1, T_2 pounds weight be the tensions in the tight and slack sides of the belt. The moment about the centre of the forces producing the rotation of the shaft B is $(T_1 - T_2) b$, and the work done per second by this couple is $(T_1 - T_2) b\omega$, or $(T_1 - T_2) v$ if v is the linear velocity of the belt. Hence the rate of working is $(T_1 - T_2) v$ foot-pounds per second, or the horse-power transmitted is $(T_1 - T_2) v/550$.

10·7. Further problems on strings and chains. Let s denote the length of a curved string or chain measured from a fixed point A on the string up to a point P, and let δs denote a short length PQ. Let the tangents at P, Q make angles ψ, $\psi + \delta\psi$ with a fixed direction. Suppose that the tension varies along

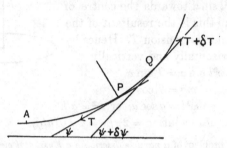

the string being T at P and $T + \delta T$ at Q. Then, no matter what are the forces that make the string assume a curved form and whether it be at rest or in motion, the effect of the tensions of the rest of the string on the element δs is represented by a force T along the tangent at P and a force $T + \delta T$ along the tangent at Q. Resolve the latter into components along the tangent and the inward normal at P, and we get along the tangent

$$- T + (T + \delta T) \cos \delta\psi$$

and along the inward normal

$$(T + \delta T) \sin \delta\psi.$$

Neglecting squares and products of δT and $\delta\psi$, these components are δT along the tangent in the sense in which s increases and $T\delta\psi$ along the inward normal.

Statical problems on strings or chains can be solved by resolving along tangent and normal for an element δs and including in the equations along with the external forces acting on the element the forces δT and $T\delta\psi$. Dynamical problems can be solved in like manner by equating the components along tangent and normal of the effective forces of the element δs to the components of the external forces together with δT and $T\delta\psi$.

For example, an **endless chain** under the action of no external force can run with uniform speed along a curve of any given form. For, the velocity v being constant, there is no acceleration along the tangent, and no external force; therefore $\delta T = 0$, or T is constant.

And if m is the mass per unit length, the mass of the element δs is $m\,\delta s$; and, if ρ is the radius of curvature, by considering the motion of this element and resolving along the inward normal we get

$$m\,\delta s v^2/\rho = T\,\delta\psi;$$

but

$$\rho = \mathrm{Lim}\ \delta s/\delta\psi,$$

therefore

$$T = mv^2.$$

If the speed of an endless chain be high the tension may be so great that external forces such as the weight of the chain are negligible in comparison, and then the chain will continue to run in the same curve for a long time, whatever form the curve may have.

10·71. Belt running on a Pulley at Uniform Speed. Suppose that the friction is limiting, i.e. just sufficient to prevent slipping of the belt on the pulley.

Let v be the velocity of the belt, $m\,\delta s$ the mass of a length δs and let a be the radius of the pulley. The normal reaction of the pulley

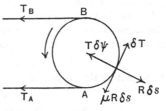

on the element δs of the belt may be denoted by $R\,\delta s$, and then the friction is $\mu R\,\delta s$ if μ is the coefficient of friction. The tensions on the element δs are equivalent to δT along the tangent and $T\,\delta\psi$ along the inward normal. Resolving along the tangent and normal we have

$$0 = \delta T - \mu R\,\delta s,$$

and

$$m\,\delta s\,\frac{v^2}{a} = T\,\delta\psi - R\,\delta s.$$

But $\delta s = a\,\delta\psi$, therefore

$$(T - mv^2)\,\delta\psi = \frac{\delta T}{\mu},$$

and, by integrating,

$$\log (T - mv^2) = \mu\psi + \text{const.},$$

or $\qquad\qquad T - mv^2 = Ce^{\mu\psi}.$

Hence if A, B are the points where the belt leaves the pulley and T_A, T_B the tensions there, and we measure ψ from the tangent at A, we have $T = T_A$ when $\psi = 0$, so that $C = T_A - mv^2$, and

$$T - mv^2 = (T_A - mv^2)e^{\mu\psi}.$$

If the belt is in contact with the pulley round a semicircle we have

$$T_B - mv^2 = (T_A - mv^2)e^{\mu\pi}.$$

If the inertia of the belt is neglected the result becomes $T_B = T_A e^{\mu\pi}$, just as if the belt were at rest.

10·8. Problems on Changing Mass. (i) *If a man on a truck running on smooth level rails is throwing out m lb. of sand per second in a direction parallel to the rails, and thus doing H ft.-lb. of work per second, prove that the velocity of the sand relative to the truck is $\sqrt{(2Hg/m)}$.*

[M. T. 1903]

Let M be the mass of the truck and its contents at time t; V, $V + \delta V$ the velocities of the same at times t, $t + \delta t$ and $-v$ the velocity of the sand relative to the truck.

In time δt a mass $m\,\delta t$ lb. of sand is thrown out with velocity $V - v$. The linear momentum is constant and, by equating the momentum at time $t + \delta t$ to its value at time t, we get

$$(M - m\,\delta t)(V + \delta V) + m\,\delta t \cdot (V - v) = MV;$$

or, neglecting the product $\delta t\,\delta V$,

$$M\,\delta V - mv\,\delta t = 0 \quad\dots\dots\dots\dots\dots\dots\dots(1).$$

Again the work done in time δt is $Hg\,\delta t$ foot-poundals and this is equal to the increase in kinetic energy.

Therefore $Hg\,\delta t = \frac{1}{2}(M - m\,\delta t)(V + \delta V)^2 + \frac{1}{2}m\,\delta t \cdot (V - v)^2 - \frac{1}{2}MV^2$

$$= MV\,\delta V - m\,\delta t \cdot Vv + \frac{1}{2}m\,\delta t \cdot v^2;$$

and by (1) this $\quad = \frac{1}{2}m\,\delta t \cdot v^2,$

so that $\qquad v = \sqrt{(2Hg/m)}.$

(ii) *A mass m of water issues per unit time from a pipe with uniform velocity u, and strikes a pail which retains it, there being no elasticity. Initially the pail is at rest, and at a subsequent instant is moving in the direction of the steam with velocity V. Prove that*

$$\frac{dV}{dt} = \frac{m(u - V)^3}{Mu^2},$$

and that the loss of energy up to this instant is

$$\tfrac{1}{2} Mu V,$$

where M is the mass of the pail, and gravity is omitted from consideration.

[S. 1910]

Let V denote the velocity of the pail and M' the mass of water in it at time t.

If α denotes the cross-section and ρ the density of the stream of water, since the velocity is u the length of the stream that emerges in unit time is u, the volume is $u\alpha$ and the mass $\rho u\alpha$. Therefore $\rho u\alpha = m$. But when the pail has velocity V the velocity of the stream relative to it is $u - V$ and therefore the length of stream entering the pail in time δt is $(u - V)\,\delta t$ and its mass is $\rho\,(u - V)\,\alpha\,\delta t = \dfrac{m(u - V)}{u}\,\delta t$, and its momentum is $m(u - V)\,\delta t$. But the momentum of the pail and its contents at time t is $(M + M')\,V$ and in time δt this increases to $(M + M' + \delta M')\,(V + \delta V)$.

The increase must be equal to the momentum of the water that enters the pail, by conservation of linear momentum.

Therefore $\qquad (M + M')\,\delta V + V\delta M' = m(u - V)\,\delta t$ ....(1).

But $\delta M'$ is the mass of water added in time δt, so that

$$\delta M' = \frac{m(u - V)}{u}\,\delta t \dots\dots\dots\dots\dots\dots\dots(2).$$

Substituting in (1) we get

$$(M + M')\,\delta V = \frac{m(u - V)^2}{u}\,\delta t \dots\dots\dots\dots\dots\dots(3).$$

Eliminate δt from (2) and (3) and we get

$$\frac{\delta M'}{M + M'} = \frac{\delta V}{u - V}\,;$$

therefore $\qquad\qquad (M + M')\,(u - V) = \text{const.}$

$$= Mu \dots\dots\dots\dots\dots\dots\dots(4),$$

because initially $M' = 0$ and $V = 0$.

Dividing (3) by (4) gives the result

$$\frac{dV}{dt} = \frac{m(u - V)^3}{Mu^2}\,.$$

Again, the loss of energy up to this point is

$$\tfrac{1}{2} M'u^2 - \tfrac{1}{2}(M + M')\,V^2,$$

and on substituting the value of M' from (4) this gives

$$\tfrac{1}{2} Mu V.$$

EXAMPLES

1. Two weights W, W' balance on any system of pulleys with vertical strings. If a weight w be attached to W, shew that it will descend with acceleration

$$g \Big/ \left[1 + \frac{W(W+W')}{W'w} \right],$$

neglecting the inertia of the pulleys. [S. 1918]

2. A light string passes over two smooth pulleys in the same horizontal line and carries masses m_1, m_2 at its extremities and a smooth ring of mass m_3 free to slide on the string between the pulleys. If all parts of the string hang vertically, prove that the ring will remain at rest provided

$$m_3(m_1+m_2) = 4m_1m_2,$$

and that in that case the acceleration of the centre of gravity of the three masses is

$$\frac{(m_1-m_2)^2}{(m_1+m_2)(m_1+m_2+m_3)}g. \qquad \text{[S. 1900]}$$

3. A mass m lying on a smooth horizontal table is attached to a string which after passing over the edge of the table hangs in a loop on which a heavy smooth ring of mass M is threaded and then passes over a smooth fixed pulley and supports a mass m'. If the free portions of the string are vertical and the whole system lies in a vertical plane, determine the tension of the string, and shew that the mass M will remain at rest provided that

$$2/M = 1/m + 1/m'. \qquad \text{[S. 1924]}$$

4. Two particles of masses m and $3m$ are connected by a fine string passing over a fixed smooth pulley. The system starts from rest and the heavier particle, after falling 8 feet, impinges on a fixed inelastic support. Find the velocity with which it is next jerked off the support; and shew that the system finally comes to rest 3 seconds from the beginning of the motion. [S. 1917]

5. A light string passes over a fixed smooth pulley and carries at one end a mass $6m$, and at the other a smooth pulley of mass $3m$ over which passes a second light string carrying masses $2m$ and m at its ends. Assuming that the system moves from rest, obtain expressions for the velocities and accelerations of the movable pulley and the masses. Use this system to verify the principle of the conservation of energy, and explain why it does not illustrate the principle of the conservation of linear momentum. What distribution of masses between the various parts of the system would cause it to do so? [S. 1924]

6. Two equal particles A and B are connected by a light inextensible string of length a which is stretched at full length perpendicular to the edge of the table. The particle A is drawn just over the edge of the table and is then released from rest in this position. Describe the nature of the subsequent motion and shew that after B leaves the table the centre of inertia of the two particles describes a parabola of latus rectum $a/2$.

[S. 1917]

7. From a gun of mass M which can recoil freely on a horizontal platform a shell of mass m is fired with velocity v, the elevation of the gun being α. Determine the direction in which the shell is moving when it leaves the gun, and shew that, if the shell strikes at right angles the plane that passes through the point of projection and is inclined to the horizontal at an angle β, then

$$\tan \alpha = M(\cot \beta + 2 \tan \beta)/(M+m).$$
[S. 1924]

8. A particle of mass 2 lb. is placed on the smooth face of an inclined plane of mass 7 lb. and slope 30°, which is free to slide on a smooth horizontal plane in a direction perpendicular to its edge. Shew that if the system starts from rest the particle will slide down a distance of 15 feet along the face of the plane in 1·25 seconds. [S. 1923]

9. A smooth wedge of mass M and angle α is free to move on a smooth horizontal plane in a direction perpendicular to its edge. A particle of mass m is projected directly up the face of the wedge with velocity V. Prove that it returns to the point on the wedge from which it was projected after a time

$$2V(M+m\sin^2\alpha)/\{(m+M)g\sin\alpha\}.$$
Also find the pressure between the particle and the wedge at any time.
[S. 1917]

10. A smooth wedge of inclination α is placed on a horizontal table; a string, to the extremities of which masses m, m' are attached, passes round a smooth peg which projects from the upper face of the wedge, the particles being in contact with the same face of the wedge. Prove that, if M be the mass of the wedge, its acceleration is

$$\frac{(m-m')^2 g \sin\alpha\cos\alpha}{M(m+m')+(m-m')^2\sin^2\alpha+4mm'},$$

assuming that the motion is in planes of greatest slope. [S. 1899]

11. A particle of mass m is placed on the smooth slant surface of a wedge of mass M and angle α. The wedge moves (edge foremost) along a smooth horizontal plane under the action of a constant horizontal force P perpendicular to its edge, all the motion being parallel to the same vertical plane. Prove that the acceleration of the wedge is $\dfrac{P-mg\sin\alpha\cos\alpha}{M+m\sin^2\alpha}$, and find the pressure between the particle and the wedge. [S. 1907]

12. A particle of mass m slides down the rough inclined face of a wedge of mass M and inclination α, which is free to move on a smooth horizontal plane. Shew that the time of describing any distance from rest is less than the time taken when the wedge is fixed in the ratio

$$\left\{1 - \frac{m \cos \alpha (\cos \alpha + \mu \sin \alpha)}{M+m}\right\}^{\frac{1}{2}} : 1. \qquad \text{[S. 1917]}$$

13. A smooth wedge of mass M resting on a horizontal plane is subject to smooth constraints so that it can only move along the plane in a direction at right angles to the intersections of its slant faces with the plane. A particle of mass m is moving along a face of the wedge which is inclined to the horizontal at an angle α so that the component velocity of the particle perpendicular to the line of greatest slope is u. Shew that the wedge moves with constant acceleration and that the path of the particle on the surface of the wedge is a parabola of latus rectum

$$\frac{2u^2}{g} \frac{M+m\sin^2\alpha}{(M+m)\sin\alpha}.$$

Verify that the principle of conservation of energy holds good in this case.

Solve the same problem when the wedge, instead of being free to move, is made to move in the same direction as before with constant velocity V.
                                                                    [S. 1925]

14. Two particles of mass M and m $(M>m)$ are placed on the two smooth faces of a light wedge which rests on a smooth horizontal plane. The faces of the wedge are inclined to the horizontal at angles α and β, respectively. If the system starts from rest, shew that the smaller particle will move up the face on which it is placed if

$$\tan \beta < \frac{M \sin \alpha \cos \alpha}{M \sin^2\alpha + m}. \qquad \text{[S. 1915]}$$

15. A particle of mass m is placed on the inclined face of a wedge of mass M which rests on a rough horizontal table. Prove that, if the particle slides down, the wedge will begin to move provided that

$$\frac{m}{M} > \frac{\cos \lambda \sin \lambda'}{\cos \alpha \sin (\alpha - \lambda - \lambda')},$$

where α is the inclination of the face of the wedge to the horizontal, λ is the angle of friction for the particle and the wedge, and λ' is the angle of friction for the wedge and the table.   [S. 1924]

16. A tube ABC of mass m is bent at right angles at B. The part AB is horizontal and slides freely through two fixed rings; the part BC is vertical. Particles P, Q, each of mass m, move without friction in AB, BC, and are connected by a string passing over a smooth pulley of negligible mass at B. The system is released from rest Apply the principles of

momentum and energy to shew that, when Q has fallen a distance y from its initial position, its vertical velocity is $\sqrt{(6gy/5)}$.

Shew that the vertical and horizontal components of the acceleration of Q are $3g/5$ and $g/5$. [M. T. 1922]

17. A smooth straight tube BAC is bent at A and is fixed in a vertical plane so that AB, AC make angles α, β on opposite sides of the vertical. A heavy uniform string PAQ in the tube is slightly displaced from the position of equilibrium; shew that when a length x has passed over A in the direction of P both the velocity and the acceleration vary as x, and that the tension of the string at A varies as $(p+x)(q-x)$, where p, q are the lengths AP, AQ in the position of equilibrium.

Shew also that the resultant vertical pressure on the tube is

$$W\left(1 - \frac{x^2}{pq}\cos\alpha\cos\beta\right),$$

where W is the weight of the string. [S. 1917]

18. A light string $ABCDE$, whose middle point is C, passes through smooth rings B, D which are fixed in a horizontal plane at a distance $2a$ apart. To each of the points A, C, E is attached a mass m. Initially C is held at rest at O, the middle point of BD, and is then set free. Shew that C will come instantaneously to rest when $OC = 4a/3$. [The total length of the string is greater than $10a/3$.]

Shew that when C has fallen through $3a/4$ from O, its velocity is

$$\sqrt{(25ag/86)}.$$ [M. T. 1922]

19. Two particles, masses M and m $(M > m)$, are attached to the ends of a string, length $2l$, which passes over a smooth peg at a height l above a smooth plane inclined at an angle α to the vertical. The particles are initially held at rest on the plane at the point vertically below the peg, M being below m. Prove that, if the particles are released, m will oscillate through a vertical distance $2M(M-m)l/(m^2\sec^2\alpha - M^2)$, provided that $\tan^2\alpha$ is greater than $(3M+m)(M-m)/m^2$. [S. 1923]

20. A ring of mass m slides on a smooth vertical rod; attached to the ring is a light string passing over a smooth peg distant a from the rod, and at the other end of the string is a mass M $(>m)$. The ring is held on a level with the peg and released: shew that it first comes to rest after falling a distance

$$\frac{2mMa}{M^2-m^2}.$$ [S. 1915]

21. A mass m is suspended at the lower end of a vertical elastic wire of mass m and length L, suspended at its upper end. The system is caused to execute small vertical oscillations. Assuming that the wire can be treated as uniformly stretched throughout the motion, shew that the

kinetic energy of the system is $\frac{1}{2}\dot{x}^2(M+\frac{1}{3}m)$, where x is the displacement of the mass M from its equilibrium position. Hence shew that the time of a complete oscillation is $2\pi\sqrt{\{kL(M+\frac{1}{3}m)\}}$, where k is such that unit force produces an extension of k units in unit length of the wire.   [M. T. 1921]

22. Two equal particles A, B attached to the ends of a light string of length a are placed on a smooth horizontal table with the string AB perpendicular to the edge of the table and B hanging just over the edge. The system is released from rest in this position. Prove that when first the string is horizontal the distance of B from the vertical through the edge of the table is $\frac{1}{4}a(\pi-2)$, and find the tension in the string.   [S. 1923]

23. A horizontal bar AB of length a is made to rotate with a constant angular velocity ω about a vertical axis through the end B. If a particle is attached to A by a string of length l, the string makes an angle θ with the vertical when the motion is steady. Prove that

$$l\cos\theta + a\cot\theta = g/\omega^2.   \text{[M. T. 1914]}$$

24. A particle of mass m' is attached by a light inextensible string of length l to a ring of mass m free to slide on a smooth horizontal rod. Initially the two masses are held with the string taut along the rod and they are then set free. Prove that the greatest angular velocity of the string is

$$\{2g(m+m')/lm\}^{\frac{1}{2}}.$$

Also shew that the time of a small oscillation about the vertical is

$$2\pi\{lm/g(m+m')\}^{\frac{1}{2}}.   \text{[Coll. Exam. 1914]}$$

25. Two particles of masses $7m$ and $3m$ are fastened to the ends A, B respectively of a weightless rigid rod, 15 ft. long, which is freely hinged at a point O, 5 ft. from A_1: if the rod is just disturbed from its position of unstable equilibrium, prove that the velocity with which A will pass through its position of stable equilibrium is $\frac{1}{15}\sqrt{380g}$.   [S. 1910]

26. Two particles, each of mass m, are joined by a rod of negligible mass and of length a. One particle rests on a smooth horizontal plane and the other is vertically above it. The upper particle is given a small displacement so that the rod begins to fall. Shew that when its inclination to the vertical is 60° the velocity of the lower particle is $\sqrt{(ga/14)}$.   [M. T. 1926]

27. A heavy ring of mass m slides on a fixed smooth vertical rod and is attached to a fine string which passes over a smooth peg distant a from the rod and then after passing through a smooth ring of mass M is tied to the peg. Prove that, if m is dropped from the point in the rod in the same horizontal line as the peg, then, provided $M>2m$, it will oscillate through a distance $4mMa/(M^2-4m^2)$ and find the greatest velocity of m.   [S. 1924]

EXAMPLES 137

28. The driving wheel of a motor cycle has a diameter of 28 ins.; the belt pulling on the driving wheel has a diameter of 20 ins. If 3 H.P. is the rate of working when the cycle is going 20 miles an hour, and in order that the belt may not slip the tension on one side may not exceed $2\frac{1}{2}$ times that on the other, shew that the tensions are $131\frac{1}{4}$ lb. wt. and $52\frac{1}{2}$ lb. wt.

[S. 1914]

29. Power is delivered by an engine through ropes passing round the flywheel of the engine and over pulleys in the mill. If the flywheel is 30 ft. in diameter and turns at 90 revolutions per minute, determine how many ropes of $1\frac{1}{2}$ inch diameter will be required to transmit 2000 horse-power. Assume the tension on the tight side to be twice the tension on the slack side, and allow not more than 300 pounds per square inch pull in the ropes. [M. T. 1918]

30. A belt-driven pulley of diameter 2 feet transmits 10 horse-power when running at 240 revolutions per minute. Shew that if the belt is just on the point of slipping, and subtends an arc of 180° of the pulley, the biggest tension in the belt is 306 lb. nearly. The coefficient of friction between belt and pulley can be taken at 0·4. [M. T. 1920]

31. A machine gun of mass M contains a mass M' of bullets which it discharges at the rate m units of mass per unit time, V being the velocity of the bullets relative to the ground. Shew that, if μ be the coefficient of friction between the gun and the ground, the whole time of recoil of the gun will be

$$(2mV - \mu g M') M'/2\mu g m M.$$ [S. 1921]

32. A horizontal rod of mass M is movable along its length, and its motion is controlled by a light spring which exerts a restoring force Ex when the rod is displaced through a distance x. A spider of mass m stands on the rod, and everything is initially at rest. The spider then runs a distance a along the rod, and then stops, his velocity relative to the rod being constant and equal to u. Shew that the total energy of the system after the run is

$$\frac{2m^2u^2}{M+m} \sin^2 \left(\frac{a}{2u} \sqrt{\frac{E}{M+m}} \right),$$

and find the amplitude of the final motion. [S. 1925]

33. A machine gun of mass M stands on a horizontal plane and contains shot of mass M'. The shot is fired at the rate of mass m per unit of time with velocity u relative to the ground. If the coefficient of sliding friction between the gun and the plane is μ, shew that the velocity of the gun backward by the time the mass M' is fired is

$$\frac{M'}{M} u - \frac{(M+M')^2 - M^2}{2mM} \mu g.$$ [M. T. 1919]

ANSWERS

22. Half the weight of either particle.   29. 30.

Chapter XI

IMPULSIVE MOTION

11·1. The state of rest or motion of a body sometimes undergoes an apparently instantaneous change owing to the sudden application of a force which acts for a very short time only. For example—a ball struck by a bat, or the collision of two billiard balls.

In such cases it is not possible to measure the *rate* of change of momentum because a finite change of momentum takes place in an infinitesimal interval of time.

In **4·2** we saw that the change of momentum produced by a variable force X acting from time $t = t_1$ to $t = t_2$ is $\int_{t_1}^{t_2} X \, dt$. Now it is possible for the force to increase and at the same time the interval $t_2 - t_1$ to decrease in such a way that the integral tends to a finite limit, although we have no means of measuring the exact value of X at any instant during the interval. It is usual to call such a force an **Impulsive Force** or **Impulse** and measure it by the change of momentum it produces, i.e. we measure it by its 'impulse' as defined in **4·2**.

It is to be noted that an impulsive force or impulse is not of the same physical dimensions as 'force.' The latter is of dimensions $\mathbf{MLT^{-2}}$ and the former of dimensions $\mathbf{MLT^{-1}}$ (**4·8**).

11·2. Equations of Motion for Impulsive Forces. The equations of motion for a system of particles acted upon by finite forces were found in **9·2** to be

$$\Sigma m\ddot{x} = \Sigma X, \quad \Sigma m\ddot{y} = \Sigma Y,$$

and

$$\frac{d}{dt} \Sigma m \, (x\dot{y} - y\dot{x}) = \Sigma \, (xY - yX).$$

If we integrate these equations with respect to t through an interval from t_0 to t, we get

$$\Sigma m\dot{x} - \Sigma m\dot{x}_0 = \Sigma \int_{t_0}^{t} X \, dt,$$

$$\Sigma m\dot{y} - \Sigma m\dot{y}_0 = \Sigma \int_{t_0}^{t} Y \, dt,$$

and

$$\Sigma m \, (x\dot{y} - y\dot{x}) - \Sigma m \, (x\dot{y}_0 - y\dot{x}_0) = \Sigma \left\{ x \int_{t_0}^{t} Y dt - y \int_{t_0}^{t} X dt \right\},$$

where $\dot{x}_0$, $\dot{y}_0$ denote the values of $\dot{x}$, $\dot{y}$ at time t_0.

Now if we suppose that we are concerned with impulsive forces as defined in **11·1**, then $\int_{t_0}^{t} X dt$ and $\int_{t_0}^{t} Y dt$ are the measures of the components of the impulse and may be denoted by P, Q respectively. Hence the last three equations may be written

$$\begin{rcases} \Sigma m\dot{x} - \Sigma m\dot{x}_0 = \Sigma P \\ \Sigma m\dot{y} - \Sigma m\dot{y}_0 = \Sigma Q \end{rcases} \quad \dots\dots\dots\dots(1),$$

and $\qquad \Sigma m \, (x\dot{y} - y\dot{x}) - \Sigma m \, (x\dot{y}_0 - y\dot{x}_0) = \Sigma \, (xQ - yP) \ \dots(2).$

These equations express the facts that

(1) *the instantaneous increase in the linear momentum in any direction is equal to the sum of the externally applied impulsive forces in that direction; and*

(2) *the instantaneous increase in the moment of momentum about any axis is equal to the sum of the moments about that axis of the externally applied impulsive forces.*

Further we observe that with the notation of **9·3** equations (1) may be written

$$M \, (\dot{\bar{x}} - \dot{\bar{x}}_0) = \Sigma P,$$

and $\qquad\qquad M \, (\dot{\bar{y}} - \dot{\bar{y}}_0) = \Sigma Q.$

Also the equations confirm the principles of conservation of linear and angular momentum, in that if there be a direction in which the external impulsive forces have zero component there is no change of momentum in that direction; and if there be an axis about which the external impulsive forces have zero moment there is no change of moment of momentum about that axis.

It is to be observed that, in writing down equations for the instantaneous change of motion produced by impulsive forces, all finite forces such as weight are to be neglected, because if F is a finite force then $\int_{t_0}^{t} F dt$ vanishes when the interval $t - t_0$ tends to zero.

11·3. Impact of Smooth Spheres. The problem of the motion of smooth spheres after impact is determined partly by the principle of conservation of linear momentum and partly by an experimental law due to Newton, viz. '*The relative velocity of the spheres along the line of centres immediately after impact is − e times the relative velocity before impact,*' where *e* is a constant depending on the substances of which the spheres are composed, called the *coefficient of restitution*. For hard substances like steel or ivory *e* is nearly unity, but for a soft substance it is small. When a substance is described as *perfectly elastic* it is to be understood that $e = 1$, and, when inelastic, $e = 0$.

The value of *e* may be determined experimentally by suspending two spheres by equal fine threads, so that when at rest the spheres are in contact and the threads vertical. Then one of the spheres is drawn back and released so as to strike the other. From **8·2** the square of the velocity of either sphere at the lowest point of its path, i.e. at striking, 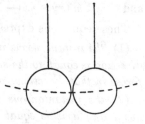 is proportional to the vertical height through which it has descended or to which it ascends. Hence by measuring the vertical heights all the velocities immediately before and after impact can be found, and so *e* is determined.

11·31. Direct Impact. Let m, m' be the masses of the two spheres, u, u' their velocities before impact and v, v' their velocities after impact and let the motion be along the line of centres.

The momentum in the line of motion is unaltered by the impact, so that

$$mv + m'v' = mu + m'u' \quad \dots\dots\dots\dots\dots(1);$$

and, by Newton's rule,

$$v - v' = - e(u - u') \quad \dots\dots\dots\dots\dots(2).$$

These equations determine the velocities after impact, namely

$$v = \frac{mu + m'u' - em'(u - u')}{m + m'},$$

and

$$v' = \frac{mu + m'u' + em(u - u')}{m + m'}.$$

The impulse between the spheres which reduces the velocity of the first from u to v is $m(u-v)$, which is equal to

$$\frac{(1+e)\,mm'\,(u-u')}{m+m'}:$$

and this is $(1+e)$ times what the impulse would be if the coefficient of restitution were zero.

11·32. Poisson's Hypothesis. This hypothesis is that when the bodies come into contact there is a short interval in which they undergo compression followed by another short interval in which the original shape is restored. At the instant of greatest compression the bodies have a common velocity along the line of centres, and the impulsive pressure between the bodies during restitution is less than the impulsive pressure during compression in the ratio $e:1$. Now if the bodies were inelastic (i.e. if $e=0$) they would acquire a common velocity

$$(mu+m'u')/(m+m')$$

and there would be no restitution, so the hypothesis is that when the bodies are elastic the whole impulsive pressure during impact is $(1+e)$ times the impulse during compression, and from **11·31** this is in accordance with Newton's rule.

In fact we may deduce Newton's rule from this hypothesis. If with the notation of the last article we denote by I the impulse during compression and by U the common velocity at the instant of greatest compression, then

$$m(u-U)=I, \text{ and } m'(U-u')=I.$$

Eliminate U, then

$$u-u'=\frac{I}{m}+\frac{I}{m'}, \quad \ldots\ldots\ldots\ldots\ldots\ldots(1).$$

But, by hypothesis, the whole impulse $=(1+e)\,I$, therefore

$$m(u-v)=(1+e)\,I, \text{ and } m'(v'-u')=(1+e)\,I.$$

Whence we get

$$u-u'-(v-v')=(1+e)\,I\left(\frac{1}{m}+\frac{1}{m'}\right),$$

and from (1) $=(1+e)(u-u');$

therefore $v-v'=-e(u-u'),$

which is Newton's rule.

11·33. Oblique Impact. When the directions of motion of
the spheres are not along the line
of centres their velocities can be
resolved along and perpendicular
to the line of centres. We may
use the symbols of **11·31** to denote
the velocities along the line of
centres, and they will satisfy the

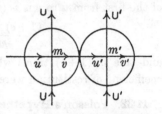

equations of that article. When we consider the motion of
either sphere at right angles to the line of centres, we see that,
since the spheres are smooth, there is no impulsive action that
can affect the velocity in this direction. Hence if U, U' denote
the component velocities at right angles to the line of centres
before impact, they also denote the components in the same
directions after impact.

11·34. Kinetic Energy lost by Impact. In general there is
a loss of kinetic energy in the cases we are considering. We
need only consider the case of direct impact, because, in oblique
impact, the square of the velocity being the sum of the squares
of its perpendicular components and components at right angles
to the line of centres being unaltered by impact, it follows that
it is only the velocity components in the line of centres that
can effect a change in kinetic energy. With the notation of
11·31 and using the theorem of **9·5** we write the kinetic energy
before impact in the form

$$\tfrac{1}{2}mu^2 + \tfrac{1}{2}m'u'^2 = \tfrac{1}{2}(m+m')\left(\frac{mu+m'u'}{m+m'}\right)^2$$
$$+ \tfrac{1}{2}m\left(u - \frac{mu+m'u'}{m+m'}\right)^2 + \tfrac{1}{2}m'\left(u' - \frac{mu+m'u'}{m+m'}\right)^2,$$

where $(mu+m'u')/(m+m')$ is the velocity of the centre of
gravity and the terms in the last two brackets represent the
velocities of the spheres relative to the centre of gravity. This
reduces to

$$\tfrac{1}{2}\frac{(mu+m'u')^2}{m+m'} + \tfrac{1}{2}\frac{mm'(u-u')^2}{m+m'}.$$

Similarly the kinetic energy after impact

$$= \tfrac{1}{2}\frac{(mv+m'v')^2}{m+m'} + \tfrac{1}{2}\frac{mm'(v-v')^2}{m+m'}.$$

But $$mv + m'v' = mu + m'u',$$

and $$v - v' = -e(u - u');$$

therefore there is a loss of kinetic energy equal to

$$\tfrac{1}{2} mm' (1 - e^2) (u - u')^2 / (m + m'),$$

or $(1 - e^2)$ times the kinetic energy of the spheres relative to the centre of gravity before impact.

11·35. Generalization of Newton's Rule. For impacts of bodies other than spheres we may generalize Newton's rule and say that *the velocities, before and after impact, of the points of two bodies that come into contact resolved along the common normal at the point of contact are in the ratio* $1 : -e$.

Thus suppose a sphere moving with velocity u strikes a smooth plane in a direction making an angle θ with the normal to the plane, and that it rebounds with velocity v making an angle ϕ with the normal.

The above rule makes

$$v \cos \phi = eu \cos \theta;$$

and since the velocity parallel to the plane is unaltered

$$v \sin \phi = u \sin \theta,$$

therefore $$\cot \phi = e \cot \theta.$$

So that the angles of incidence and reflection of such a sphere impinging on the plane are governed by this law of reflection.

11·36. Example. *Two equal spheres of mass m' are suspended by vertical strings so that they are in contact with their centres at the same level. A third equal sphere of mass m falls vertically and strikes the other two simultaneously so that their centres at the instant of impact form an equilateral triangle in a vertical plane. If u is the velocity of m just before impact, find the velocities just after impact and the impulsive tension of the strings.*

After impact the spheres of mass m' begin to move horizontally because of the constraints of the strings; let v be the velocity of either; and let u' be the velocity of m, which by symmetry must be vertical. Let I denote the impulse between the upper and either of

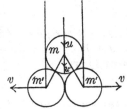

the lower spheres, T the impulsive tension of the strings and e the coefficient of restitution.

By Newton's rule

$$u' \cos 30° - v \cos 60° = - eu \cos 30°,$$

or
$$\sqrt{3} u' - v = - \sqrt{3} eu \dots\dots\dots\dots\dots\dots\dots\dots(1).$$

Resolving vertically for m,
$$m (u - u') = 2I \cos 30° \dots\dots\dots\dots\dots\dots(2),$$
and horizontally for m', $m'v = I \cos 60°$ $\dots\dots\dots\dots\dots\dots\dots(3).$

Eliminating I from (2) and (3), we get
$$m (u - u') = 2 \sqrt{3} m'v;$$
and from (1) $= 6m' (u' + eu),$
therefore $u' = u (m - 6em')/(m + 6m');$
and from (1) $v = \sqrt{3} um (1 + e)/(m + 6m').$

Again, resolving vertically for m', we get
$$T = I \cos 30° = \sqrt{3} m'v;$$
and from (3) $= 3mm'u (1 + e)/(m + 6m').$

11·4. As further examples of impulsive action we take the following:

Examples. (i) *A body of mass $m_1 + m_2$ is split into two parts of masses m_1 and m_2 by an internal explosion which generates kinetic energy E. Shew that if after explosion the parts move in the same line as before, their relative speed is*

$$\sqrt{\{2E (m_1 + m_2)/m_1 m_2\}}.$$ [M. T. 1902]

Let u be the speed of the body before explosion, and v_1, v_2 the speeds of m_1, m_2 after explosion.

There is no change in linear momentum, therefore
$$m_1 v_1 + m_2 v_2 = (m_1 + m_2) u.$$

The kinetic energy is increased by E, therefore
$$\tfrac{1}{2} m_1 v_1^2 + \tfrac{1}{2} m_2 v_2^2 = \tfrac{1}{2} (m_1 + m_2) u^2 + E.$$

Eliminate u, therefore
$$(m_1 v_1^2 + m_2 v_2^2) (m_1 + m_2) = (m_1 v_1 + m_2 v_2)^2 + 2 (m_1 + m_2) E,$$
or $m_1 m_2 (v_1 - v_2)^2 = 2 (m_1 + m_2) E,$
and $v_1 \sim v_2 = \sqrt{\{2E (m_1 + m_2)/m_1 m_2\}}.$

(ii) *A light rigid rod ABC has three particles each of mass m attached to it at A, B, C. The rod is struck by a blow P at right angles to it at a point distant from A equal to BC. Prove that the kinetic energy set up is*

$$\frac{1}{2} \frac{P^2}{m} \frac{a^2 - ab + b^2}{a^2 + ab + b^2},$$

where $AB = a$, and $BC = b$.

The motion of the rod is completely defined by its angular velocity ω and the linear velocity u of any one point, say C. The velocity of B is then $u+b\omega$ and the velocity of A is

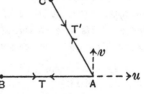

$$u+(a+b)\,\omega.$$

To find the two unknown quantities u, ω we resolve at right angles to the rod and equate the total momentum to the impulse, i.e.

$$mu+m\,(u+b\omega)+m\,(u+\overline{a+b}\omega)=P,$$

or

$$3u+\omega\,(a+2b)=P/m \quad\ldots\ldots\ldots\ldots\ldots\ldots\ldots\ldots(1).$$

We also take moments, and the point O, at which the blow is applied, is the most convenient point about which to take moments, because P has no moment about O, so that the total moment of momentum about O is zero. Therefore

$$m\{u+(a+b)\,\omega\}\,b-m\,(u+b\omega)\,(a-b)-mua=0,$$

or

$$u\,(a-b)=b^2\omega \quad\ldots\ldots\ldots\ldots\ldots\ldots\ldots\ldots\ldots(2).$$

From (1) and (2)

$$\frac{u}{b^2}=\frac{\omega}{a-b}=\frac{P}{m}\cdot\frac{1}{3b^2+(a+2b)\,(a-b)}$$

$$=\frac{P}{m}\cdot\frac{1}{a^2+ab+b^2}\,.$$

And the kinetic energy created by the blow

$$=\tfrac{1}{2}m\,(u+\overline{a+b}\omega)^2+\tfrac{1}{2}m\,(u+b\omega)^2+\tfrac{1}{2}mu^2$$

$$=\frac{1}{2}\frac{P^2}{m}\frac{a^2-ab+b^2}{a^2+ab+b^2}\,.$$

(iii) *Three equal particles A, B, C of mass m are placed on a smooth horizontal plane. A is joined to B and C by light threads AB, AC and the angle BAC is $60°$. An impulse I is applied to A in the direction BA. Find the initial velocities of the particles and shew that A begins to move in a direction making an angle $\tan^{-1}\sqrt{3}/7$ with BA.* [S. 1924]

Let u, v denote the initial velocities of A along and at right angles to BA, and let T, T' denote the impulsive tensions in AB, AC.

Since B and A are joined by an inextensible string their velocities along BA must be the same, therefore the velocity of B is u along BA. For a similar reason the velocity of C is $u\cos 60°-v\cos 30°$ or $\tfrac{1}{2}(u-\sqrt{3}v)$ along CA.

Now, writing down the equations of impulsive motion of the particles in turn, we have for A

$$mu=I-T-\tfrac{1}{2}T',\quad mv=\frac{\sqrt{3}}{2}\,T'\,;$$

for B $\qquad\qquad mu=T,$

and for C $\qquad \tfrac{1}{2}m\,(u-\sqrt{3}v)=T'.$

Eliminating T' gives

$$\tfrac{1}{2}(u-\sqrt{3}v)=\frac{2}{\sqrt{3}}v, \quad \text{or} \quad \sqrt{3}u=7v.$$

Then by eliminating T and T'' from the first equation we get

$$2u+v/\sqrt{3}=I/m,$$

so that $u=7I/15m,$

and $v=\sqrt{3}\,I/15m.$

Also the direction of motion of A makes with BA an angle

$$\tan^{-1}v/u=\tan^{-1}\sqrt{3}/7.$$

11·5. Kinetic Energy created by Impulses. Consider a system of particles of masses m_1, m_2, m_3, ... moving with velocities whose components parallel to the axes Ox, Oy are (u_1, v_1), (u_2, v_2), (u_3, v_3), ... Let them be acted upon by impulses whose components in the same directions are (X_1, Y_1), (X_2, Y_2), (X_3, Y_3), ..., including the 'internal' as well as the 'external' impulses on every particle; and let the velocities be changed by the impulses to (u_1', v_1'), (u_2', v_2'), (u_3', v_3'), ... Then the equations of impulsive motion of the particles are

$$m_1(u_1'-u_1)=X_1, \quad m_1(v_1'-v_1)=Y_1,$$
$$m_2(u_2'-u_2)=X_2, \quad m_2(v_2'-v_2)=Y_2,$$
$$m_3(u_3'-u_3)=X_3, \quad m_3(v_3'-v_3)=Y_3,$$
$$\cdots\cdots\cdots\cdots\cdots\cdots\cdots$$

Multiply each of these equations by half the sum of the velocity components that it involves and add them all together, and we get

$$\tfrac{1}{2}\Sigma m(u'^2+v'^2)-\tfrac{1}{2}\Sigma m(u^2+v^2)=\tfrac{1}{2}\Sigma X(u+u')+\tfrac{1}{2}\Sigma Y(v+v').$$

This equation expresses the fact that *the change of kinetic energy created by a set of simultaneous impulses is the sum of the products of each impulse into the mean of the velocities of its point of application resolved in its direction.*

11·51. Let us consider how the last theorem is illustrated by examples (i) and (ii) of **11·4**.

(i) Here the impulse I is an internal one which increases the velocity of (say) m_1 from u to v_1, and decreases the velocity of m_2 from u to v_2, acting in opposite directions on the two masses.

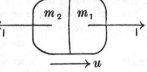

Hence $I=m_1(v_1-u)=m_2(u-v_2).$

And, by the last theorem, the energy created by the simultaneous impulses is

$$E = \tfrac{1}{2} I (v_1 + u) + \tfrac{1}{2} I (-v_2 - u)$$
$$= \tfrac{1}{2} m_1 (v_1{}^2 - u^2) + \tfrac{1}{2} m_2 (v_2{}^2 - u^2),$$

as is otherwise obvious.

(ii) The velocity of the point O is zero initially, and after the blow it is v say, where

$$v = u + a\omega = \frac{P}{m} \cdot \frac{a^2 - ab + b^2}{a^2 + ab + b^2}.$$

Therefore the kinetic energy created by the blow

$$= \tfrac{1}{2} P v = \frac{1}{2} \frac{P^2}{m} \cdot \frac{a^2 - ab + b^2}{a^2 + ab + b^2}.$$

11·6. Elasticity and Impulses. It is to be observed that an elastic or deformable body as distinct from a rigid body yields to an impulsive action and does not begin to offer resistance until a finite deformation has taken place. Thus if an impulse is applied to an *elastic* string there is no impulsive tension set up in the string, and it is not until a finite elongation has taken place that a finite tension is produced. For example in **11·36**, if the strings were elastic, the initial velocities of the spheres m' would not be horizontal but along the lines joining the centre of m to their centres.

This property of an elastic string is also illustrated in the following example.

11·61. Example. *Two particles of masses m_1 and m_2 are connected by a fine elastic string of natural length l and modulus of elasticity λ. They are placed on a smooth horizontal table at a distance l apart, and equal impulses I in opposite directions act simultaneously on them in the line of the string so as to extend it. Prove that the greatest extension of the string in the ensuing motion is*

$$I \{(m_1 + m_2)\, l / m_1 m_2 \lambda\}^{\frac{1}{2}},$$

and that this value is attained in time

$$\tfrac{1}{2} \pi \{m_1 m_2 l / (m_1 + m_2) \lambda\}^{\frac{1}{2}}. \qquad \text{[M. T. 1916]}$$

Let x, y denote displacements of m_1, m_2 in opposite directions in time t, and let $x + y = z$. The tension at time t is therefore $\lambda z / l$.

The equations of motion of the particles are therefore

$$m_1 \ddot{x} = -\lambda z / l, \quad \text{and} \quad m_2 \ddot{y} = -\lambda z / l.$$

By adding we get

$$\ddot{x} + \ddot{y} = -\frac{\lambda z}{l} \left(\frac{1}{m_1} + \frac{1}{m_2} \right),$$

or

$$\ddot{z} = -\frac{\lambda z (m_1 + m_2)}{l m_1 m_2} \quad \dots\dots\dots\dots\dots\dots\dots\dots(1).$$

This represents a simple harmonic motion so long as the string remains stretched and the time during which the velocity $\dot{z}$ changes from its maximum to zero is one quarter of the period, i.e.

$$\tfrac{1}{2}\pi \, \{m_1 m_2 l / (m_1 + m_2) \, \lambda\}^{\frac{1}{2}}.$$

Again the initial velocities of m_1 and m_2 are I/m_1 and I/m_2, since the string being elastic has no impulsive tension. Therefore the kinetic energy created

$$= \tfrac{1}{2} m_1 \left(\frac{I}{m_1}\right)^2 + \tfrac{1}{2} m_2 \left(\frac{I}{m_2}\right)^2$$

$$= \tfrac{1}{2} I^2 \, \frac{m_1 + m_2}{m_1 m_2}.$$

Now when the string has its greatest extension the particles are momentarily at rest and the kinetic energy has been converted into the potential energy of the stretched string. Hence if z is the greatest extension and T the greatest tension, the work done in stretching

$$= \tfrac{1}{2} Tz = \tfrac{1}{2} \lambda \frac{z^2}{l};$$

and by equating this to the kinetic energy we get

$$z = I \, \{(m_1 + m_2) \, l / m_1 m_2 \lambda\}^{\frac{1}{2}}.$$

We might also obtain this result by multiplying equation (1) by $2\dot{z}$ and integrating, and so finding the velocity $\dot{z}$ in terms of z; then find the value of z for which $\dot{z}$ vanishes.

EXAMPLES

1. A smooth sphere impinges on another one at rest; after the collision their directions of motion are at right angles. Shew that if they are assumed perfectly elastic, their masses must be equal.   [S. 1926]

2. Two smooth elastic spheres (coefficient of restitution e) impinge obliquely in any manner; one of them being initially at rest, it is found that the angle between their subsequent directions of motion is constant. Find the ratio of the masses of the spheres, and the angle.   [S. 1921]

3. Two equal spheres of mass $9m$ are at rest and another sphere of mass m is moving along their line of centres between them. How many collisions will there be if the spheres are perfectly elastic?   [S. 1923]

4. n equal perfectly elastic spheres move with given velocities under no forces in the same straight line. Shew that the final velocities of the spheres depend only on their order and not on their initial distances apart.   [S. 1926]

5. Shew that if a smooth sphere of mass m_1 collides with another smooth sphere of mass m_2 at rest, and is deflected through an angle θ from its former path, the sphere of mass m_2 being set in motion in a direction ϕ with the former path of m_1, then $\tan\theta=\dfrac{m_2\sin 2\phi}{m_1-m_2\cos 2\phi}$, both spheres being perfectly elastic.   [S. 1925]

6. A body of mass m rests on a smooth table. Another of mass M moving with velocity V collides with it. Both are perfectly elastic and smooth and no rotations are set up by the collision. The body m is driven in a direction at angle θ to the previous line of the body M's motion. Shew that its velocity is $\dfrac{2M}{M+m}\,V\cos\theta$.   [S. 1921]

7. A sphere collides obliquely with another sphere of equal mass which is initially at rest, both spheres being smooth and perfectly elastic. Shew that their paths after collision are at right angles.

The centres of two such spheres, B, C, each of 3 cms. radius, are at E, F, where $EF=16$ cms. An equal sphere A is projected with velocity u at right angles to EF, and strikes first B and then C. Its final path is at right angles to EF. Find the point of contact between A and B. Shew that
$$v_A=9u/25,\quad v_B=20u/25,\quad v_C=12u/25,$$
where v_A, v_B, v_C are the final velocities.   [M. T. 1922]

8. The masses of three spheres A, B, C are $7m$, $7m$, m; their coefficient of restitution is unity. Their centres are in a straight line and C lies between A and B. Initially A and B are at rest and C is given a velocity along the line of centres in the direction of A. Shew that it strikes A twice and B once, and that the final velocities of A, B, C are proportional to 21, 12, 1.   [M. T. 1915]

9. Two equal marbles, A and B, lie on a smooth horizontal circular groove at opposite ends of a diameter. A is projected along the groove and at the end of time t impinges on B; shew that the second impact will occur after a further time $2t/e$, where e is the coefficient of restitution.   [M. T. 1921]

10. Two smooth and perfectly elastic spheres of masses 1 and 4 respectively are initially at rest under no forces. The more massive sphere is then projected in such a direction as to strike the other sphere and rebound. Prove that the direction of motion of the more massive sphere cannot be deflected by the collision through an angle greater than $14° 29'$.   [M. T. 1925]

11. Two equal pendulums OP, CQ, of length l, are suspended from two points O, C in a horizontal line, such that when the bobs are hanging at rest they are just in contact. The bob P is projected horizontally with velocity $\sqrt{gl}$ from the point at height l vertically above O, and strikes the

bob Q which was previously hanging at rest. Shew that the string of Q will become slack before Q reaches its highest point, if the coefficient of restitution lies between $\sqrt{1 \cdot 6} - 1$ and unity.   [M. T. 1924]

12. Two small spheres A and B of equal mass m, the coefficient of restitution between which is e, are suspended in contact by two equal vertical strings so that the line of centres is horizontal. The sphere A is drawn aside through a small distance and allowed to fall back and collide with the other, its velocity on impact being u. Shew that all subsequent impacts occur when the spheres are in the same position as for the first impact, and that the velocity of the sphere A immediately after the third impact is $\frac{1}{2} u (1 - e^3)$. Shew that the kinetic energy of the system tends to the value $\frac{1}{4} m u^2$.   [M. T. 1926]

13. Two imperfectly elastic particles of equal mass, whose coefficient of restitution is e, are suspended from the same point by light strings of equal length. One particle is drawn aside a *small* distance x_0 and then released. Shew that, between the nth and $(n+1)$th impacts, the particle originally drawn aside swings through a distance $\frac{1}{2} \{1 + (-e)^n\} x_0$ on one side of the vertical through the point of suspension.   [S. 1923]

14. A spherical ball of mass m suspended by a string from a fixed point is at rest, and another spherical ball of mass m' which is falling vertically with velocity u impinges on it so that the line joining the centres of the balls makes an angle α with the vertical. Prove that the loss of energy

$$= \frac{1}{2} (1 - e^2) \, m m' u^2 \cos^2 \alpha / (m + m' \sin^2 \alpha),$$

where e is the coefficient of restitution.   [S. 1924]

15. Three equal similar spheres of mass m' are suspended by equal vertical threads so that their centres are at the corners of an equilateral triangle in a horizontal plane. A fourth smooth sphere, of mass m, falls vertically so as to strike the other spheres simultaneously. Determine the velocities immediately after impact, having given the velocity u of the fourth sphere and the angle θ which the lines of centres make with the vertical at the instant of striking.   [S. 1925]

16. A steel ball is released from rest and falls upon a fixed steel anvil and rebounds, the coefficient of restitution being 0·9. The lowest point of the ball is initially at a distance of one foot above the anvil. Find the position and velocity of the ball half a second after its release.

Shew that the ball finally comes to rest on the anvil 4·75 seconds after its release and that the total distance travelled is $9\frac{10}{19}$ feet.   [M. T. 1914]

17. A spherical particle is let fall vertically under gravity and after describing a distance h impinges at a point A on a smooth plane inclined at an angle α to the horizontal. Shew that the particle ceases to rebound from the plane when it reaches a point B such that

$$AB = 4he \sin \alpha / (1 - e)^2,$$

where e is the coefficient of restitution.   [M. T. 1920]

18. Two smooth spheres of equal mass, whose centres are moving with equal speeds in the same plane, collide in such a way that at the moment of collision the line of centres makes an angle $\frac{1}{2}\pi - \epsilon$ with the direction bisecting the angle α between the velocities before impact. Shew that after impact the velocities are inclined at an angle $\tan^{-1}(\cos 2\epsilon \tan \alpha)$, the collision being perfectly elastic.   [M. T. 1923]

19. Two particles of masses M and m are connected by a fine inextensible string passing over a fixed smooth pulley, and the motion of the heavier particle, M, is limited by a fixed horizontal inelastic plane, on which it can impinge. The system starts from rest with M at a given height above the plane; shew that the successive heights of M at which it comes to instantaneous rest form a geometrical progression of ratio $\{m/(M+m)\}^2$, and that the whole time of motion is three times the interval from the beginning of the motion to the first impact on the plane.   [M. T. 1916]

20. A bucket of mass m_1 is joined to a counterpoise of mass m_2 by a light string hanging over a smooth pulley. A ball of mass m is dropped into the bucket. Shew that the ball will come to rest in the bucket at a time $\dfrac{ev(m_1+m_2)}{(1-e)m_2 g}$ after the first impact, where v is the velocity of the ball relative to the bucket immediately before the first impact, and e is the coefficient of restitution.

Shew that the sum of the upward momentum of the system on one side of the pulley, and the downward momentum of that on the other side, increases at a uniform rate, and determine this rate. Hence or otherwise shew that the velocity of the system so soon as the ball has come to rest in the bucket is

$$u + \frac{mm_2 + e(mm_1 + m_1{}^2 - m_2{}^2)}{(1-e)m_2(m+m_1+m_2)}\, v,$$

where u is the downward velocity of the bucket immediately before the first impact.   [S. 1925]

21. A railway truck is at rest at the foot of an incline of 1 in 70. A second railway truck of equal weight starts from rest at a point 1000 feet up the incline, and runs down under gravity. The trucks collide at the foot of the incline, the coefficient of restitution being $\frac{1}{2}$. Find how far each truck travels along the level, the frictional resistances for each truck being 16 lb. wt. per ton, both on the incline and on the level. Where the incline meets the level, the rails are slightly curved, each in a vertical plane, so that there is no vertical impact, and at the instant of collision both trucks are on the level.   [S. 1925]

22. A heavy elastic particle is projected from a point O at the foot of an inclined plane of inclination α to the horizon. The plane through the direction of projection normal to the inclined plane meets the inclined plane in a line OA which makes an angle ϕ with the line of greatest slope

and the direction of projection makes an angle θ with OA. Find equations to determine the position of the particle after any number of rebounds and shew that the particle will just have ceased to rebound when it again reaches the foot of the plane if

$$\tan\theta\tan\alpha=(1-e)\cos\phi,$$

where e is the coefficient of restitution.   [S. 1926]

23. Shew that after an elastic collision $(e=1)$ between two equal smooth spheres, one of which is initially at rest but free to move in any direction, the directions of motion of the two spheres are at right angles.

Shew further that if the mass of the resting sphere is greater in the ratio $1+\epsilon:1$ (ϵ small), then the angle between the directions of motion will exceed a right angle by $\frac{1}{2}\epsilon\tan\phi$ approximately, where ϕ is the deflection of the moving sphere.   [M. T. 1927]

24. A heavy perfectly elastic particle is dropped from a point P on the inside surface of a smooth sphere. Prove that the second point of impact on the surface of the sphere will be in the same horizontal plane as the first if the angular distance of P from the highest point of the sphere is

$$\cos^{-1}\{(2^{\frac{1}{2}}+1)^{\frac{1}{2}}/2\}.$$ [S. 1919]

25. A sphere of mass m impinges directly on a sphere of mass m' at rest on a smooth table. The second sphere then strikes a vertical cushion at right angles to its path. Shew that there will be no further impact of the spheres if $m(1+e'+ee')<em'$; where e,e' are the coefficients of restitution between the spheres and between the sphere and the cushion.   [S. 1921]

26. Two equal flat scale pans are suspended by an inextensible string passing over a smooth pulley so that each remains horizontal. An elastic sphere falls vertically and when its velocity is u it strikes one of the scale pans and rebounds vertically. Shew that the sphere takes the same time to come to rest on the scale pan as it would if the scale pan were fixed.   [S. 1924]

27. A ball is projected on a pocketless billiard table. Shew that, if the effect of friction and rotation be neglected, it will travel always parallel to one of two fixed directions so long as it strikes the four cushions in order: and that the velocity is decreased in the ratio $e^2:1$ after each complete circuit, e being the coefficient of restitution.   [S. 1922]

28. A bullet of 0·1 lb. weight is fired with a speed of 2200 feet per second into the middle of a block of wood of 30 lb. weight, which is at rest but free to move. Find the speed of the block and bullet afterwards, and the loss of kinetic energy in foot-pounds. What becomes of this energy?
                [M. T. 1917]

29. A bullet of mass m is fired into a block of wood of mass M, which is free to move on a smooth horizontal table, and penetrates it to a depth a. Shew that, at the instant when the bullet comes to rest relative to the block, the block has moved through a distance $ma/(M+m)$, the stress between the bullet and the block being assumed constant, so long as there is any relative motion. [S. 1911]

30. A, B, C are three equal particles attached to a light inextensible string at equal intervals a. The system is placed on a smooth horizontal plane with the three particles in a straight line. A blow P is applied to the middle one B in a direction perpendicular to the string. Describe the nature of the subsequent motion, and shew that the angular velocity of AB or BC is $P/ma\,(2+\cos\theta)^{\frac{1}{2}}$ where θ is the angle ABC and m the mass of a particle. [S. 1913]

31. Four particles, each of mass m, are connected by equal inextensible strings of length a and lie on a table at the corners of a rhombus the sides of which are formed by the strings. One of the particles receives a blow P along the diagonal outwards. Prove that the angular velocities of the strings after the blow are equal to $P\sin\alpha/2ma$, where $2\alpha\left(\alpha<\dfrac{\pi}{4}\right)$ is the angle of the rhombus at the particle which is struck. [S. 1926]

32. Three particles A, B, C each of the same mass rest on a smooth table at the corners of an equilateral triangle; AB and BC being tight inextensible strings. A is given a velocity v in the direction CB. Shew that when the string AB again tightens C starts off with velocity $\frac{1}{15}v$. [S. 1921]

33. Three equal particles A, B, C connected by inelastic strings AB, BC of length a lie at rest with the strings in a straight line on a smooth horizontal table. B is projected with velocity V at right angles to AB. Shew that the particles A and C afterwards collide with relative velocity

$$2V/\sqrt{3}.$$

If the coefficient of restitution is e, find the velocities of the three particles when the string is again straight. [S. 1927]

34. Three masses m_1, m_2 and m_3 lie at the points A, B and C upon a smooth horizontal table; A and B, B and C are connected by light inextensible strings, and the angle ABC is obtuse. An impulse I is applied to the mass m_3 in the direction BC: find the initial velocities of the masses and shew that the mass m_2 begins to move in a direction making an angle θ with AB where

$$m_2\tan\theta+(m_1+m_2)\tan B=0. \qquad\text{[S. 1926]}$$

35. Four equal masses are attached at equal distances A, B, C, D at points on a light string, and so placed that $\angle ABC = \angle BCD = 120°$, and the various parts of the string are straight; an impulse I is given to the mass at A in the direction BA, shew that the impulsive tension in AB is $\frac{15}{28}I$.

[S. 1910]

36. A light string passing through a smooth ring at O on a smooth horizontal table has particles each of mass m attached to its ends A and B. Initially the particles lie on the table with the portions of string OA, OB straight and $OA = OB$. An impulse P is applied to the particle A in a direction making 60° with OA. Prove that when B reaches O its velocity is $P\sqrt{22}/8m$.

[S. 1923]

37. Two equal particles connected by an elastic string which is at its natural length and straight, lie on a smooth table, the string being such that the weight of either particle would produce in it an extension a. Prove that if one particle is projected with velocity u directly away from the other, each will have travelled a distance $\pi u \sqrt{\dfrac{a}{8g}}$ when the string first returns to its natural length.

[S. 1925]

38. A man of mass m is standing in a lift of mass M, which is descending with velocity V; the counterpoise being of mass $M + m$. Suddenly the man jumps with an impulse which would raise him to a height h if he were jumping from the ground. Calculate the velocities of the man and the lift, immediately after the impulse; and find also their subsequent accelerations. Deduce that the height in the lift to which he jumps is $h(M+m)/(M+\frac{1}{2}m)$.

39. A particle of mass M is at rest at a point A on a smooth horizontal plane. It is attracted towards another point B in the plane by a force proportional to its distance from B. At the instant at which it is released it is given an impulse I in the direction AB. If $AB = x_0$, and if the initial value of the attractive force is P_0, shew that the particle reaches B after a time

$$\sqrt{\frac{Mx_0}{P_0}} \tan^{-1} \frac{\sqrt{MP_0 x_0}}{I}.$$

Find an expression for the rate at which the force attracting the particle towards B is doing work, and shew that it has a maximum value

$$\frac{1}{2}\sqrt{\frac{P_0}{Mx_0}}\left(P_0 x_0 + \frac{I^2}{M}\right). \qquad \text{[M. T. 1926]}$$

ANSWERS

2. $e:1$; $\frac{1}{2}\pi$.   3. 3.

15. $\dfrac{u(m\sin^2\theta - 3m'e\cos^2\theta)}{m\sin^2\theta + 3m'\cos^2\theta}$, $\dfrac{mu(1+e)\sin\theta\cos\theta}{m\sin^2\theta + 3m'\cos^2\theta}$.

16. ·8 ft. above anvil, downward velocity ·8 ft.

21. 160 ft.; 360 ft.   28. 7·3 f.s.; 7537·3 ft.-lb.; converted into heat.

33. $\frac{1}{3}V(1+e)$, $\frac{1}{3}V(1-2e)$, $\frac{1}{3}V(1+e)$.

Chapter XII

POLAR COORDINATES. ORBITS

12·1. Velocity and Acceleration in Polar Coordinates.
Let the position of a point P be
defined by its distance r from a fixed
origin O and the angle θ that OP
makes with a fixed axis Ox. The
cartesian coordinates (x, y) of P are
connected with the polar coordinates
(r, θ) by the relations $x = r \cos \theta$,
$y = r \sin \theta$.

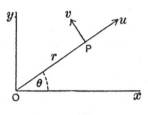

Let u, v denote the components of velocity of P in the direction
OP and at right angles to OP in the sense in which θ increases.
The resultant of the components u, v is also the resultant of the
components $\dot{x}, \dot{y}$. Therefore by resolving parallel to Ox and Oy
we get

$$u \cos \theta - v \sin \theta = \dot{x} = \frac{d}{dt}(r \cos \theta) = \dot{r} \cos \theta - r\dot{\theta} \sin \theta,$$

and $\quad u \sin \theta + v \cos \theta = \dot{y} = \dfrac{d}{dt}(r \sin \theta) = \dot{r} \sin \theta + r\dot{\theta} \cos \theta.$

Solving these equations for u and v clearly gives

$$u = \dot{r} \quad \text{and} \quad v = r\dot{\theta};$$

and these are the polar components of velocity.

In like manner if f_1, f_2 denote the components of acceleration
along and at right angles to OP, since these have the same
resultant as $\ddot{x}$ and $\ddot{y}$, we get

$$f_1 \cos \theta - f_2 \sin \theta = \ddot{x} = \frac{d^2}{dt^2}(r \cos \theta)$$
$$= (\ddot{r} - r\dot{\theta}^2) \cos \theta - (r\ddot{\theta} + 2\dot{r}\dot{\theta}) \sin \theta,$$

and $\quad f_1 \sin \theta + f_2 \cos \theta = \ddot{y} = \dfrac{d^2}{dt^2}(r \sin \theta)$
$$= (\ddot{r} - r\dot{\theta}^2) \sin \theta + (r\ddot{\theta} + 2\dot{r}\dot{\theta}) \cos \theta;$$

giving on solution $f_1 = \ddot{r} - r\dot{\theta}^2, f_2 = r\ddot{\theta} + 2\dot{r}\dot{\theta}$.

These components constitute a third representation of the velocity and acceleration of a point moving in a plane (see **5·12**); they are sometimes called **radial and transverse components,** and we note that the transverse component of acceleration may also be written $\frac{1}{r}\frac{d}{dt}(r^2\dot\theta)$.

12·2. Central Orbits. If a particle is describing an orbit under the action of a force directed to a fixed point

(i) *the orbit must be a plane curve;* because at any instant the particle is moving in the plane through the tangent to its path and the fixed point and the only force acting on the particle lies in this plane, therefore the particle continues to move in this plane.

(ii) *the rate of description of area by the radius vector drawn from the fixed point to the particle is constant;* for there is no force at right angles to the radius vector so that the transverse component of acceleration is zero throughout the motion, i.e. $\frac{1}{r}\frac{d}{dt}(r^2\dot\theta)=0$, therefore $r^2\dot\theta=$ constant. Since a sectorial element of area is $\frac{1}{2}r^2\delta\theta$, it follows that $r^2\dot\theta$ is twice the rate of description of area by the radius vector. It is usual to denote this constant by h; and we note that $r^2\dot\theta$ is also the moment of the velocity about the fixed point and that the following forms are equivalent:

$$h=r^2\dot\theta=pv=x\dot y-y\dot x.$$

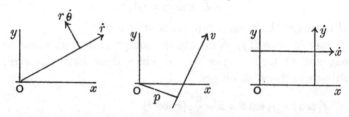

12·21. Given the Orbit and the Centre of Force to determine the Law of Force. There are two common formulae for the law of force, one for use when the orbit is given by its (r, p) equation, where r denotes the radius vector and p the perpendicular from the origin to the tangent, and the other for use when the orbit is given by its polar equation, i.e. a relation between r and θ.

(i) *Orbit given by (r, p) equation.* Let f denote the required force per unit mass, i.e. the acceleration towards the centre O.

Then, if v is the velocity, the acceleration along the inward normal is v^2/ρ, therefore by resolving along the normal

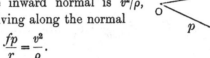

$$\frac{fp}{r} = \frac{v^2}{\rho}.$$

But $\rho = r\,dr/dp$ and $vp = h$, therefore

$$f = \frac{rh^2}{p^3\rho} = \frac{h^2}{p^3}\frac{dp}{dr} \quad\dots\dots\dots\dots\dots\dots(1).$$

(ii) *Orbit given by (r, θ) equation.* If we write u for $\dfrac{1}{r}$, and make use of the formula

$$\frac{1}{p^2} = u^2 + \left(\frac{du}{d\theta}\right)^2,$$

to be found in books on Calculus, we have

$$-\frac{2}{p^3}\frac{dp}{dr} = \frac{d}{dr}\left(\frac{1}{p^2}\right) = \frac{d}{d\theta}\left\{u^2 + \left(\frac{du}{d\theta}\right)^2\right\}\bigg/\frac{dr}{d\theta};$$

and since $\dfrac{dr}{d\theta} = -\dfrac{1}{u^2}\dfrac{du}{d\theta}$, therefore the last relation reduces to

$$\frac{1}{p^3}\frac{dp}{dr} = u^2\left\{u + \frac{d^2u}{d\theta^2}\right\},$$

and (1) may be written

$$f = h^2u^2\left\{u + \frac{d^2u}{d\theta^2}\right\} \quad\dots\dots\dots\dots\dots\dots(2).$$

Alternatively, we may start from the polar components of acceleration and write

$$\ddot{r} - r\dot{\theta}^2 = -f, \quad \frac{1}{r}\frac{d}{dt}(r^2\dot{\theta}) = 0;$$

and the latter giving $r^2\dot{\theta} = h$, it follows that

$$\frac{d}{dt} = \frac{h}{r^2}\frac{d}{d\theta} = hu^2\frac{d}{d\theta},$$

where $u = \dfrac{1}{r}$.

Substituting this operator for $\dfrac{d}{dt}$ and $\dfrac{1}{u}$ for r in the expression $\ddot{r} - r\dot{\theta}^2$, we get

$$f = -hu^2 \frac{d}{d\theta}\left\{ hu^2 \frac{d}{d\theta}\left(\frac{1}{u}\right)\right\} + h^2 u^3$$

$$= hu^2 \frac{d}{d\theta}\left\{ hu^2 \frac{1}{u^2}\frac{du}{d\theta}\right\} + h^2 u^3,$$

or $\qquad\qquad f = h^2 u^2 \left\{\dfrac{d^2 u}{d\theta^2} + u\right\}$ ....(2),

as before.

12·22. The converse problem—*given the law of force to find the orbit*—is solved by substituting the given expression for f in formula (1) or (2) and then integrating the resulting equation, and provided that the orbit can be identified by its (r, p) equation it is clear that (1) is the simpler formula to use as the orbit will be found by a single integration.

12·3. Circular Orbits. Since a particle describing a circle with uniform velocity v has a constant acceleration v^2/a towards the centre, where a is the radius, it follows that a particle can describe a circle under any constant force per unit mass f tending to the centre provided that it is projected at right angles to the radius with velocity $\sqrt{(af)}$.

Circle with any internal point O as centre of force. Let a be the radius and c the distance of O from the centre C. Then if P is any point on the circle and p the perpendicular from O to the tangent at P, and $OP = r$, it is easy to see from the triangle OCP that

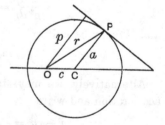

$$c^2 = r^2 + a^2 - 2ap,$$

so that $\qquad dp/dr = r/a.$

Therefore the force towards O under which the particle would describe the circle is

$$f = \frac{h^2}{p^3}\frac{dp}{dr} = \frac{h^2 r}{p^3 a},$$

the velocity at P being given by $vp = h$, or $v = \dfrac{h}{p}$.

In the special case in which O is on the circumference, $c = a$ and $r^2 = 2ap$, so that

$$f = \frac{8a^2h^2}{r^5}, \text{ and } v = \frac{2ah}{r^2};$$

or

$$f = \frac{\mu}{r^5}, \text{ and } v = \frac{\sqrt{\tfrac{1}{2}\mu}}{r^2}.$$

Conversely a particle projected from a point P at a distance r from O with velocity $\sqrt{\tfrac{1}{2}\mu}/r^2$ under the action of a force μ/r^5 per unit mass towards O will describe a circle passing through O, and the position of the centre of the circle and therefore also its radius depends only on the direction of projection; for we construct the position of C by drawing PC at right angles to the direction of projection and making the angle POC equal to OPC.

12·4. Elliptic Orbit. Force directed to the Centre. Let P be a point on the ellipse whose centre is C and semi-axes a, b.

If p is the central perpendicular on the tangent at P and CD is the radius conjugate to CP, from the properties of the ellipse, we have

$$CP^2 + CD^2 = a^2 + b^2, \text{ and } p.CD = ab.$$

Hence, if $CP = r$, we have

$$\frac{a^2b^2}{p^2} = a^2 + b^2 - r^2 \quad\dots\dots\dots\dots\dots\dots(1).$$

This is the (r, p) equation of the ellipse when the centre is the origin. By differentiating (1) we get

$$\frac{a^2b^2}{p^3}\frac{dp}{dr} = r.$$

Hence the force to the centre necessary for the description of the ellipse is

$$f = \frac{h^2}{p^3}\frac{dp}{dr} = \frac{h^2}{a^2b^2}r = \mu r, \text{ say,}$$

where μ is a constant h^2/a^2b^2, or where $h = ab\sqrt{\mu}$.

Again the velocity at P is given by

$$v = \frac{h}{p} = \frac{h}{ab} \cdot CD = \sqrt{\mu} \cdot CD,$$

and is parallel to CD.

Also since h is twice the rate of description of areas, therefore the time taken to describe the ellipse or the **periodic time in the orbit** = twice the area divided by h

$$= 2\pi ab/h = 2\pi/\sqrt{\mu}.$$

Let A be an end of the major axis and Q the point on the auxiliary circle corresponding to the point P on the ellipse; then if t denotes the time taken for the particle to move from A to P we have $t = (2 \text{ sector } ACP)/h.$

But the ellipse is the projection of the circle, so that

$$\text{area } ACP : \text{area } ACQ = b : a,$$

and $\text{area } ACQ = \frac{1}{2}a^2 \cdot A\hat{C}Q;$

therefore $\text{area } ACP = \frac{1}{2}ab \cdot A\hat{C}Q,$

and $t = \dfrac{ab}{h} \cdot A\hat{C}Q = A\hat{C}Q/\sqrt{\mu},$

or the angle $ACQ = \sqrt{\mu}t$. It follows that as P moves round the ellipse the corresponding point Q moves round the auxiliary circle with uniform angular velocity $\sqrt{\mu}$.

Further, the coordinates of the point P in terms of the eccentric angle $\sqrt{\mu}t$ are given by $x = a \cos \sqrt{\mu}t,\ y = b \sin \sqrt{\mu}t.$ This shews that the elliptic motion can be compounded of two simple harmonic motions along lines at right angles having the same period $2\pi/\sqrt{\mu}$ and differing in phase by a quarter of a period **7·11**.

12·41. Law of Force μr. Find the Orbit. The law of force gives $\dfrac{h^2}{p^3}\dfrac{dp}{dr} = \mu r$, and on integration

$$\frac{h^2}{p^2} = A - \mu r^2 \ldots\ldots(1).$$

But the (r, p) equation of an ellipse with the centre as origin is

$$\frac{a^2 b^2}{p^2} = a^2 + b^2 - r^2 \ldots(2),$$

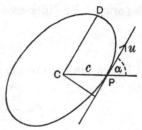

so that the orbit is in general an ellipse.

The constants involved are determined by the initial circumstances. Thus, if the particle be projected with velocity u from a point P at a distance c from the centre of force C in a direction making an angle α with CP, we have

$$h = \text{moment of velocity about } C = uc \sin \alpha \, ;$$

and since h/p always measures the velocity, therefore substituting in (1) gives $u^2 = A - \mu c^2$ or $A = u^2 + \mu c^2$.

Now by comparing (1) and (2) we get

$$a^2 + b^2 = \frac{A}{\mu} = \frac{u^2}{\mu} + c^2,$$

and

$$a^2 b^2 = \frac{h^2}{\mu} = \frac{u^2 c^2 \sin^2 \alpha}{\mu}.$$

These equations determine the lengths of the semi-axes of the ellipse in terms of the data μ, u, c, α. The inclination θ of the major axis to CP is then to be found from the polar equation of the ellipse, viz.

$$\frac{\cos^2 \theta}{a^2} + \frac{\sin^2 \theta}{b^2} = \frac{1}{r^2}.$$

12·5. Elliptic Orbit. Force directed to Focus. Let S, H be the foci of an ellipse and SY, HZ the perpendiculars to the tangent at P. To find the (r, p) equation with S as origin we assume three properties of the ellipse

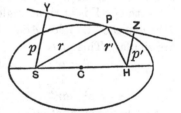

(i) $SP + HP = 2a$,

(ii) $SY . HZ = b^2$,

(iii) the tangent is equally inclined to the focal distances so that SPY, HPZ are similar triangles. Hence, if $SP = r$, $HP = r'$, $SY = p$ and $HZ = p'$, we have

$$\frac{p}{r} = \frac{p'}{r'}, \text{ therefore also} = \sqrt{\frac{pp'}{rr'}} = \frac{b}{\sqrt{\{r(2a - r)\}}}.$$

Hence $\dfrac{b}{p} = \sqrt{\left(\dfrac{2a - r}{r}\right)}$, or, squaring,

$$\frac{b^2}{p^2} = \frac{2a}{r} - 1 \quad \dotfill (1).$$

Therefore $\dfrac{b^2}{p^3}\dfrac{dp}{dr}=\dfrac{a}{r^2}$, and the force to the focus S necessary for the description of the ellipse is given by

$$f=\frac{h^2}{p^3}\frac{dp}{dr}=\frac{h^2a}{b^2}\cdot\frac{1}{r^2}=\frac{\mu}{r^2},\text{ say } \quad\ldots\ldots\ldots\ldots(2),$$

where $\mu=h^2a/b^2$ or $h^2=\mu l$, if l denotes the semi-latus rectum b^2/a.

Again the velocity $v=h/p$, therefore

$$v^2=\frac{h^2}{p^2}=\frac{h^2}{b^2}\left(\frac{2a}{r}-1\right),\text{ from (1),}$$

but $h^2=\mu b^2/a$, therefore $v^2=\dfrac{\mu}{a}\left(\dfrac{2a}{r}-1\right)$,

or $$v^2=\frac{2\mu}{r}-\frac{\mu}{a} \quad\ldots\ldots\ldots\ldots\ldots\ldots\ldots(3).$$

Again, as in **12·4**, the periodic time $=2\pi ab/h$

$$=2\pi ab/\surd\,(\mu b^2/a)$$

$$=2\pi a^{\frac{3}{2}}/\surd\mu \quad\ldots\ldots\ldots\ldots\ldots\ldots(4).$$

12·51. Parabolic Orbit. Force directed to Focus. The $(r,\ p)$ equation of a parabola is obtained from the facts that if the tangent at P meets the tangent at the vertex A in Y and S is the focus, then SY is at right angles to PY and the triangles ASY, YSP are similar. This gives

$$p^2=ar\ldots\ldots\ldots\ldots(1),$$

or $$\frac{1}{p^2}=\frac{1}{ar},$$

so that by differentiation

$$\frac{2}{p^3}\frac{dp}{dr}=\frac{1}{ar^2}.$$

Hence the force to the focus necessary for the description of the parabola is given by

$$f=\frac{h^2}{p^3}\frac{dp}{dr}=\frac{h^2}{2a}\cdot\frac{1}{r^2}=\frac{\mu}{r^2},\text{ say } \quad\ldots\ldots\ldots\ldots(2),$$

where $\mu=h^2/2a$, or $h^2=\mu\times$ semi-latus rectum.

Also, if v is the velocity at P,

$$v^2=\frac{h^2}{p^2}=\frac{2a\mu}{ar}=\frac{2\mu}{r}\ldots\ldots\ldots\ldots\ldots\ldots(3).$$

12·52. Hyperbolic Orbit. Force directed to Focus. To find the (r, p) equation of a hyperbola we assume the corresponding properties as for the ellipse, viz.:

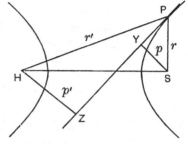

(i) $SY \cdot HZ = b^2$,

(ii) $r' - r = 2a$,

(iii) the tangent bisects the angle between the focal distances.

Hence as in **12·5** we have

$$\frac{p}{r} = \frac{p'}{r'} = \sqrt{\frac{pp'}{rr'}} = \frac{b}{\sqrt{r\,(2a + r)}},$$

so that

$$\frac{b^2}{p^2} = \frac{2a}{r} + 1 \quad \text{.....................(1)},$$

and by differentiation

$$\frac{b^2}{p^3} \frac{dp}{dr} = \frac{a}{r^2}.$$

Therefore

$$f = \frac{h^2}{p^3} \frac{dp}{dr} = \frac{h^2 a}{b^2} \cdot \frac{1}{r^2} = \frac{\mu}{r^2}, \text{ say } \quad \text{............(2)},$$

where $\mu = h^2 a/b^2$, or $h^2 = \mu l$, if l denotes the semi-latus rectum b^2/a.

In this case

$$v^2 = \frac{h^2}{p^2} = \frac{h^2}{b^2} \left(\frac{2a}{r} + 1 \right) = \frac{\mu}{a} \left(\frac{2a}{r} + 1 \right),$$

or

$$v^2 = \frac{2\mu}{r} + \frac{\mu}{a} \quad \text{.....................(3)},$$

gives the velocity.

12·53. Law of Force μ/r^2. Find the Orbit. We now have

$$\frac{h^2}{p^3} \frac{dp}{dr} = \frac{\mu}{r^2},$$

and, by integration, therefore

$$\frac{h^2}{p^2} = \frac{2\mu}{r} + C, \text{ where } C \text{ is a constant } \quad \text{.........(1).}$$

Comparing this with the (r, p) equations of the ellipse, parabola and hyperbola obtained in the last three articles, viz.:

$$\frac{b^2}{p^2} = \frac{2a}{r} - 1, \quad p^2 = ar, \quad \text{and} \quad \frac{b^2}{p^2} = \frac{2a}{r} + 1,$$

we see that the orbit required is an ellipse, parabola or hyperbola according as the constant of integration C in (1) is negative, zero, or positive.

Now h/p always denotes the velocity, so that if the particle is projected with a velocity V from a point at a distance c from the centre of force, by substituting in (1) we have

$$V^2 = \frac{2\mu}{c} + C \quad \text{......................(2),}$$

and therefore C is negative, zero or positive according as

$$V^2 <, \; = \text{ or } > 2\mu/c.$$

Hence the required orbit is an ellipse, parabola or hyperbola according as

$$V^2 <, \; = \text{ or } > 2\mu/c.$$

Also by comparing (2) of this article with (3) of the last three articles, we see that

in the ellipse $C = -\mu/a$,

in the parabola $C = 0$,

in the hyperbola $C = \mu/a$.

Therefore the semi-major axis of the ellipse or hyperbola is given in terms of the initial velocity and distance by the equations

$$V^2 = \frac{2\mu}{c} \mp \frac{\mu}{a}.$$

Also when the initial circumstances are given h, the moment of the velocity about the centre of force is known, and then in each case the semi-latus rectum $= h^2/\mu$, so that the dimensions of the orbit are completely determined. The determination of the position of the major axis is left as an exercise for the student.

12·54. Velocity Components. *When a particle describes an ellipse about a centre of force in the focus the velocity can be resolved into two components of constant magnitude, one perpendicular to the radius vector and the other perpendicular to the major axis.*

Let SP meet HZ in H', then since the tangent is equally inclined to SP, PH it is easy to prove that $HZ = ZH'$, but $HC = CS$, therefore CZ is parallel to SP.

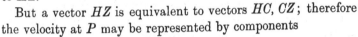

Again, the velocity at P in the ellipse is

$$\frac{h}{SY} = \frac{h}{b^2} HZ,$$

and its direction is perpendicular to HZ.

But a vector HZ is equivalent to vectors HC, CZ; therefore the velocity at P may be represented by components

$$\frac{h}{b^2} HC \text{ perpendicular to } HC,$$

and $\dfrac{h}{b^2} CZ$ perpendicular to CZ.

Since $HC = ae$ where e is the eccentricity, and $CZ = a$ and is parallel to SP therefore the velocity components are $\dfrac{hae}{b^2}$ or $\dfrac{e\mu}{h}$ perpendicular to the major axis, and $\dfrac{ha}{b^2}$ or $\dfrac{\mu}{h}$ perpendicular to SP.

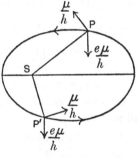

It is easy to see with the help of a figure in what sense these components must lie in order always to give a resultant in the direction of the motion of the particle, and we notice that the component $e\mu/h$ always acts in the same sense and the component μ/h is always directed in front of the radius vector SP as it revolves.

12·55. Velocity from Infinity. In connection with central orbits the phrase 'velocity from infinity' at any point of an orbit means the velocity that a particle would acquire if it moved from infinity to that particular point under the action of an attractive force in accordance with the law associated with the orbit.

Thus if f stands for acceleration towards the origin, by resolving along the tangent to the path, we get

$$v\frac{dv}{ds} = -f\frac{dr}{ds},$$

and therefore the velocity from infinity is given by

$$\tfrac{1}{2}v^2 = -\int_\infty^r f\,dr.$$

For example, if $f = \mu/r^2$,

$$v^2 = 2\mu/r,$$

so that the orbit is an ellipse, parabola or hyperbola according as the velocity at any point is less than, equal to or greater than the velocity from infinity (12·53).

12·56. The Hodograph. It is convenient here to mention the hodograph, for though it has no special connection with central orbits they afford some of the simplest illustrations of it. *If from a fixed origin O a line Oa is drawn to represent the velocity of a moving point P the locus of a is called the hodograph of the path of P.*

Referring to the figure of **5·11**, when the points P, Q are sufficiently near to one another, ab may be regarded as an elementary arc of the hodograph. It is there shewn that the acceleration of P is $\lim\limits_{\delta t \to 0} \dfrac{ab}{\delta t}$, but this is the velocity of the point a in the hodograph, so that *the velocity with which the point a describes the hodograph is a measure in magnitude and direction of the acceleration with which the point P describes its path.*

As examples of the hodograph consider the two cases of elliptic motion:

(i) when the centre of the ellipse is the centre of force, the velocity is $\sqrt{\mu}\,CD$ in magnitude and direction (12·4), therefore the hodograph is a similar ellipse;

(ii) when the focus is the centre of force, we saw (12·54) that the velocity is $\dfrac{h}{b^2}HZ$ perpendicular to HZ. But the locus of Z is the auxiliary circle, therefore in this case the hodograph is a circle.

12·6. Kepler's Laws of Planetary Motion. The following laws were announced by Kepler (1571–1630) as the result of observations of the planets:

I. *The planets describe ellipses with the sun in a focus.*

II. *The areas described by radii drawn from the sun to a planet are proportional to the times of describing them.*

III. *The squares of the periodic times are proportional to the cubes of the mean distances of the planets from the sun.*

The discovery of these laws was followed some sixty years later by Newton's enunciation of the **Law of Gravitation:**

Every particle in the universe attracts every other particle with a force which is directly proportional to the product of their masses and inversely proportional to the square of the distance between them. Thus, if m, m' denote the masses of two particles and r their distance apart, the force of attraction between them is $\gamma \dfrac{mm'}{r^2}$; where γ is a constant known as the gravitation constant, representing the attraction between two particles of unit mass at unit distance apart.

In his *Principia* Newton proved a series of propositions in Mechanics which enable us to follow a process of reasoning from Kepler's Laws to the Law of Gravitation.

Thus Newton proved that *Every body, which moves in any curved line described in a plane, and describes areas proportional to the times of describing them about a point either fixed or moving uniformly in a straight line, by radii drawn to that point, is acted on by a centripetal force tending to the same point* (*Principia*, Sect. II, Prop. II).

Hence by combining this proposition with Kepler's first and second laws, it follows that the planets, which describe ellipses about the sun in a focus, and describe areas proportional to the times of describing them by radii drawn to the sun, must be acted upon by forces directed to the sun.

Newton also proved that *If a body describes an ellipse under the action of a force tending to a focus of the ellipse, the force must vary inversely as the square of the distance* (*Principia*, Sect. III, Prop. XI).

Hence by Kepler's first law the sun must attract a planet with a force varying inversely as the square of the distance.

Then in the same section of the *Principia*, Sect. III, Props. XIV and XV, Newton shewed that *If any number of bodies revolve about a common centre, and the centripetal force vary inversely as the square of the distance, the latera recta of the orbits described will be in the duplicate ratio of the areas, which the bodies will describe in the same time by radii drawn to the centre of force.* And *On the same supposition, the squares of the periodic times in ellipses are proportional to the cubes of the major axes.* These two propositions are demonstrated on the hypothesis that there is a common law of force for all the different bodies, i.e. that if the force be denoted by μ/r^2 then μ is the same for all the bodies revolving about the common centre, and only on this hypothesis would these theorems be a logical consequence of what precedes. But Kepler's third law states as a matter of observation that for the planets the squares of the periodic times *are* proportional to the cubes of the mean distances (or major axes). The inference is that for the attraction of the sun on the planets there is a common law of force μ/r^2 per unit mass, where μ is the same constant for all the planets. And since this expression μ/r^2 is the force *per unit mass* or the acceleration imparted to the planet by the sun, therefore the whole force exerted by the sun on the planet varies as (mass of planet)$/r^2$. It is only a short step further to infer that the force is also proportional to the mass of the sun, and the law of universal gravitation is a generalization from these inferences.

Calculations based upon the law of gravitation were sufficient to enable the astronomers, Adams and Le Verrier, independently to determine the existence and position of the planet Neptune before it had been actually observed, and predictions of the return of comets have been fulfilled. In fact the law has proved adequate as a basis for dynamical astronomy generally, though there are a few known phenomena where calculations shew minute differences from the results of observation, and these have been explained and accounted for by the substitution of Einstein's theory of Relativity for the absolute Newtonian dynamics.

12·61. Necessary Modification of Kepler's Third Law.
In the preceding argument the acceleration produced by the
action of the sun on a planet is considered and the acceleration
of the sun produced by the action of the planet is ignored.
Both should be taken into account, for the observations made
are based on the motion of the planet relative to the sun and
not on an absolute motion of the planet.

Assuming that the sun and planet attract one another like
two particles, if S and P denote their masses and r their
distance apart, the force on each towards the other is $\gamma SP/r^2$,
where γ is the gravitation constant.

$$S \bullet\!\!-\!\!\rightarrow\!\!\rule{3cm}{0.4pt}\!\!\leftarrow\!\!-\bullet P$$
$$\gamma SP/r^2 \qquad\qquad \gamma SP/r^2$$

Therefore their accelerations are as shewn in the next
diagram,

$$S \bullet\!\!-\!\!\rightarrow \qquad\qquad \leftarrow\!\!-\bullet P$$
$$\gamma P/r^2 \qquad\qquad \gamma S/r^2$$

and to find the acceleration of the planet relative to the sun we
must add to the acceleration of both that of the sun reversed in
direction,

$$S \bullet \qquad\qquad\qquad \leftarrow\!\!-\bullet P$$
$$\gamma (S+P)/r^2$$

so that the acceleration of the planet relative to the sun is
$\gamma (S+P)/r^2$, and this is the μ/r^2 of our theory. Hence the
periodic time for the relative motion is

$$\frac{2\pi a^{\frac{3}{2}}}{\sqrt{\{\gamma (S+P)\}}};$$

and Kepler's third law should read 'The square of the periodic
time is proportional to the cube of the mean distance and
inversely proportional to the sum of the masses of the sun and
the planet.'

Since the mass of the largest planet, Jupiter, is less than
one-thousandth of the sun's mass it is clear that this correction
to the law is a small one.

It must be remembered that the units are such that $\gamma mm'/r^2$ is a force, so that in terms of fundamental units

$$\gamma M^2 L^{-2} = MLT^{-2},$$

therefore
$$\gamma = M^{-1}L^3T^{-2},$$

and this makes $2\pi a^{\frac{3}{2}}/\sqrt{\{\gamma (S + P)\}}$ of one dimension in time.

12·62. Examples. (i) *The eccentricity of the earth's orbit round the sun is 1/60; prove that the earth's distance from the sun exceeds the semi-axis major of the orbit during about 2 days more than half the year.*

[M. T. 1908]

If S be the sun in the focus of the ellipse $ABA'B'$, then, since $SB=$semi-axis major, therefore the distance from S of a point P on the curve exceeds the semi-axis major so long as P is on the arc $B'A'B$, and since time of describing an arc is proportional to sectorial area, therefore the time required

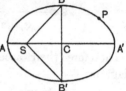

$= \dfrac{\text{area } SB'A'BS}{\pi ab}$ of a year, where a, b are the semi-axes,

$= \frac{1}{2} + \dfrac{SC \cdot CB}{\pi ab} = \frac{1}{2} + \dfrac{e}{\pi}$ of a year, where e is eccentricity,

$= \frac{1}{2}$ a year $+ \dfrac{1}{60\pi} \times 365\frac{1}{4}$ days,

$= \frac{1}{2}$ a year $+$ about 2 days.

(ii) *The greatest and least velocities of a certain planet in its orbit round the sun are 30 and 29·2 kilometres per second. Find the eccentricity of the orbit.*

[M. T. 1919]

Since the moment of the velocity about the focus S is constant, the greatest and least velocities are where the distances from S of the tangent to the path are least and greatest, i.e. at the ends A, A' of the major axis.

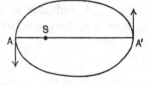

Hence $30\,SA = 29·2\,SA'$,

or $30a\,(1-e) = 29·2a\,(1+e)$,

therefore $59·2e = ·8$ or $e = 1/74$.

(iii) *The acceleration of a particle in the gravitational field of a star is μ/r^2 towards the centre of the star, where μ is a constant and r is the distance from the centre of the star. A particle starts at a great distance with velocity V, the length of the perpendicular from the centre of the star on the tangent to the initial path of the particle being p. Shew that the least distance of the particle from the centre of the star is λ, where*

$$V^2\lambda = (\mu^2 + p^2\,V^4)^{\frac{1}{2}} - \mu.$$

If $\mu=1$, $V=2$, in the system of units chosen, and if the radius of the star is 0·005, prove that the particle will strike the star if p is less than 0·05.

[M. T. 1925]

Since the particle starts from a great distance, we may assume that $V^2>2\mu/c$, where c denotes this great distance, so that by **12·53** the orbit is a
hyperbola, with the centre of the star at
the focus S.

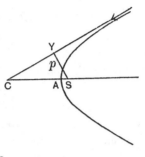

Initially the particle is moving along
the asymptote at a great distance, and p,
the perpendicular SY from the focus to
the asymptote, is known to be equal to
the semi minor axis b. This also follows
from the properties of the hyperbolic orbit;
for $V^2p^2=h^2=\mu b^2/a$ (**12·52**), and the ve-
locity v at distance r from S is given by $v^2=\dfrac{2\mu}{r}+\dfrac{\mu}{a}$, but at a great distance
we may put $r=\infty$ and $v=V$, so that $V^2=\mu/a$; therefore from above $p^2=b^2$.

Now the least distance of the particle from S is AS; therefore, if e
denotes the eccentricity,

$$\lambda=SA=CS-CA=a(e-1),$$

but $$p^2=b^2=a^2(e^2-1),\text{ and }V^2a=\mu,$$

therefore $$V^4p^2=\mu^2(e^2-1)\text{ or }\mu^2e^2=\mu^2+V^4p^2.$$

Hence $$V^2\lambda=V^2a(e-1)=\mu(e-1),$$

or $$V^2\lambda=(\mu^2+V^4p^2)^{\frac12}-\mu.$$

Substituting the numerical values $\mu=1$ and $V=2$ we have

$$4\lambda=(1+16p^2)^{\frac12}-1;$$

and the particle will strike the star if it would pass so that its least
distance from the centre is less than the radius, i.e. if $\lambda<·005$, or if
$4\lambda<·02$. Therefore the necessary condition is

$$(1+16p^2)^{\frac12}-1<·02,$$

or $$1+16p^2<(1·02)^2,$$

or $$p<·05.$$

12·7. Use of u, θ Formulae. So far we have only made use
of the formula which expresses the law of force in terms of
r, p, and this method is adequate for most simple problems.
We shall now make use of the alternative formula

$$f=h^2u^2\left(\frac{d^2u}{d\theta^2}+u\right).$$

Law of force μu^2. Find the orbit.

Since $f = \mu u^2$, therefore

$$\frac{d^2 u}{d\theta^2} + u = \frac{\mu}{h^2}.$$

Putting $u' = u - \mu/h^2$, the equation becomes

$$\frac{d^2 u'}{d\theta^2} + u' = 0,$$

the integral of which is $u' = A \cos(\theta - \varpi)$ (see **1·7** (16)), or

$$u = \frac{\mu}{h^2} + A \cos(\theta - \varpi),$$

where A, ϖ are constants of integration. This equation may be written

$$\frac{h^2/\mu}{r} = 1 + \frac{A h^2}{\mu} \cos(\theta - \varpi) \quad \ldots\ldots\ldots\ldots(1),$$

which we recognize as the polar equation of a conic with the focus as origin and semi-latus rectum h^2/μ, since the typical form of the equation of a conic is

$$\frac{l}{r} = 1 + e \cos \theta.$$

The constants of integration can be found when the circumstances of projection are known and the form of the conic can be determined, but the details are not so simple as in **12·53**, where the same problem is solved by the (r, p) equation.

12·71. Inverse Cube. Let the force to the centre be μu^3, then we have

$$\frac{d^2 u}{d\theta^2} + u = \frac{\mu}{h^2} u.$$

The solution of this equation is different in form according as

$$\mu/h^2 > = \text{ or } < 1.$$

First let $\mu/h^2 - 1 = n^2$, then the equation is

$$\frac{d^2 u}{d\theta^2} - n^2 u = 0,$$

the solution of which is $u = A e^{n\theta} + B e^{-n\theta}$ (**1·7** (14)), where the constants A, B depend on the circumstances of projection; if these are such that either A or B is zero the path is an equiangular spiral.

Secondly, when $\mu/h^2 = 1$, the equation is $\dfrac{d^2u}{d\theta^2} = 0$, the solution of which is $u = A\theta + B$, the curve known as the reciprocal spiral.

In the third case, putting $1 - \mu/h^2 = n^2$, the equation is

$$\frac{d^2u}{d\theta^2} + n^2u = 0,$$

and the solution is

$$u = A \cos n\theta + B \sin n\theta \quad\ldots\ldots\ldots\ldots 1·7\ (16),$$

a curve with infinite branches.

12·72. Apses and Apsidal Distances. An apse is a point in a central orbit at which the normal to the curve passes through the centre of force, and the length of the radius vector at such a point is called the apsidal distance.

Whenever the central force is a function of the distance, the velocity and the inclination of the path to the radius vector are also functions of the distance; for by resolving along the tangent to the path we get

$$v\frac{dv}{ds} = -f\frac{dr}{ds};$$

therefore $\tfrac{1}{2}v^2 = C - \int f dr,$

which is a function of r alone.

Again, if ϕ is the angle between the tangent and radius vector

$$\sin \phi = p/r, \text{ but } pv = h,$$

therefore $\sin \phi = h/rv,$

which is also a function of r.

It follows that at all points in the orbit which are at the same distance from O the velocity v and the inclination ϕ have the same values. Hence an apse line, i.e. a line from the centre to an apse must divide the orbit symmetrically, as is also obvious from the fact that if particles were projected from an apse in opposite directions at right angles to the apse line they would describe the same curve in opposite senses.

There can therefore be only two different apsidal distances though there may be any number of apses; for if A, B, C, D, etc. are consecutive apses, by symmetry about OB we have $OA = OC$, by symmetry about OC we have $OB = OD$ and so on.

The analytical condition for an apse, i.e. that the tangent to the curve must be at right angles to the radius vector, is that $du/d\theta$ vanishes and changes sign as θ increases through the value that indicates the position of an apse.

In the motion of a planet about the sun there is a single apse line, namely the major axis of the orbit. The apse nearer to the sun is called **perihelion** and the further apse is called **aphelion**. In like manner if the relative motion of the earth and sun be ascribed to the sun the apse nearer to the earth is called **perigee** and the further apse is called **apogee**.

12·73. Example. When it is required to find an orbit for a more complicated law of force it is generally necessary to make use of the (u, θ) equation, and we shall illustrate the general method of procedure by other examples.

If the law of force is $5\mu u^3 + 8\mu c^2 u^5$, and the particle is projected from an apse at the distance c with velocity $3\sqrt{\mu}/c$; to prove that the orbit is $r = c\cos 2\theta/3$.

We have
$$h^2 \left(\frac{d^2 u}{d\theta^2} + u \right) = 5\mu u + 8\mu c^2 u^3.$$

Multiply by $2du/d\theta$ and integrate and we get
$$h^2 \left\{ \left(\frac{du}{d\theta} \right)^2 + u^2 \right\} = 5\mu u^2 + 4\mu c^2 u^4 + C \quad \dots\dots\dots\dots(1).$$

To determine the constant C we substitute the initial values and we may do this either by observing that since $\left(\frac{du}{d\theta} \right)^2 + u^2 = \frac{1}{p^2}$, therefore the left-hand side of the equation (1) is h^2/p^2, i.e. the square of the velocity; or by observing that at the apse $du/d\theta = 0$ and $u = \frac{1}{c}$ and that h, the moment of the velocity, is $3\sqrt{\mu}$. Either process gives $C = 0$, and then, since $h^2 = 9\mu$, the equation reduces to
$$9 \left(\frac{du}{d\theta} \right)^2 = 4 \left(c^2 u^4 - u^2 \right).$$

This gives
$$\frac{du}{\sqrt{(c^2 u^4 - u^2)}} = \tfrac{2}{3} d\theta \quad \dots\dots\dots\dots\dots\dots\dots\dots(2),$$

or, putting $u = 1/r$,
$$\frac{-dr}{\sqrt{(c^2 - r^2)}} = \tfrac{2}{3} d\theta$$

so that
$$\cos^{-1} \frac{r}{c} = \tfrac{2}{3}\theta + \alpha,$$

i.e. $r = c \cos(\frac{2}{3}\theta + a)$, and if we measure θ from the apse line then $r = c$ when $\theta = 0$, therefore $a = 0$ and $r = c \cos 2\theta/3$.

It should be remarked that on taking the square root in (2) above, there is a choice of signs, and that whichever sign be chosen we get the same equation for the path.

12·74. Einstein's Law of Gravitation. One of the small discrepancies unexplained by the law of the inverse square is a slight continuous change in the position of the apse line of the planet Mercury spoken of as 'the advance of perihelion.'

We have seen in **12·7** that the Newtonian differential equation of the orbit is

$$\frac{d^2u}{d\theta^2} + u = \frac{\mu}{h^2} \quad\ldots\ldots\ldots\ldots\ldots\ldots(1),$$

the solution of which as in (1) of **12·7** is of the form

$$u = \frac{\mu}{h^2}\{1 + e \cos(\theta - \varpi)\} \quad\ldots\ldots\ldots\ldots(2);$$

and since this makes $du/d\theta = 0$ when $\theta = \varpi$, therefore ϖ is the angular coordinate of perihelion.

According to Einstein's Law of Gravitation the differential equation of the orbit is

$$\frac{d^2u}{d\theta^2} + u = \frac{\mu}{h^2} + 3\mu u^2 \ldots\ldots\ldots\ldots\ldots\ldots(3).$$

The term $3\mu u^2$ is small, for its ratio to μ/h^2 is $3h^2u^2$, i.e. three times the square of the transverse velocity measured in terms of unit velocity, which in Einstein's theory is taken to be the velocity of light. For applications in the solar system this ratio is of order 10^{-8}. Hence Einstein's theory adds a small term $3\mu u^2$ to the right hand side of equation (1). Equation (2) is therefore a first approximation to the solution of equation (3), when the small term on the right is neglected. In order to obtain a second approximation to the solution we next substitute the value of u given by (2) in the small term on the right of (3) getting

$$\frac{d^2u}{d\theta^2} + u = \frac{\mu}{h^2} + \frac{3\mu^3}{h^4} + \frac{6\mu^3}{h^4} e \cos(\theta - \varpi) + \frac{3\mu^3 e^2}{2h^4}\{1 + \cos 2(\theta - \varpi)\}.$$

The only one of the additional terms that gives appreciable

effects is the one that contains $\cos(\theta - \varpi)$, and neglecting the others we have

$$\frac{d^2u}{d\theta^2} + u = \frac{\mu}{h^2} + \frac{6\mu^3}{h^4}\,e\cos(\theta - \varpi) \quad\ldots\ldots\ldots\ldots(4).$$

Now it is easily verified that a particular integral of the equation

$$\frac{d^2u}{d\theta^2} + u = A\cos(\theta - \varpi)$$

is

$$u = \tfrac{1}{2}A\theta\sin(\theta - \varpi).$$

Hence the solution of (4) is obtained from the solution of (1) by adding to it a term

$$\frac{3\mu^3}{h^4}\,e\theta\sin(\theta - \varpi),$$

whence we get from (2)

$$u = \frac{\mu}{h^2}\{1 + e\cos(\theta - \varpi)\} + \frac{3\mu^3}{h^4}\,e\theta\sin(\theta - \varpi) \quad\ldots(5).$$

And this is of the form

$$u = \frac{\mu}{h^2}\{1 + e\cos(\theta - \varpi - \delta\varpi)\} \quad\ldots\ldots\ldots\ldots(6),$$

provided that $\delta\varpi = \dfrac{3\mu^2}{h^2}\,\theta$ and $(\delta\varpi)^2$ is neglected.

Now ϖ is the angular coordinate of perihelion and θ is the corresponding angular coordinate of the planet; and (6) shews that at any instant the planet may be considered to be moving in an ellipse but that the ellipse is not fixed, for the coordinate ϖ which defines the position of the apse line is undergoing a continuous change proportional to the angular motion of the planet in the ratio

$$\frac{\delta\varpi}{\theta} = \frac{3\mu^2}{h^2} = \frac{3\mu}{a\,(1 - e^2)},$$

since $h^2 = \mu l = \mu a\,(1 - e^2)$.

This calculated motion of the apse line in the case of the planet Mercury accords closely with the results of observation.

12·741. Example. *Shew that the orbit is a conic when the law of force is μu^2, and find the angular advance of the apse line in one period when the velocity in the orbit is less than the velocity from infinity and the central acceleration contains a small extra term αu^3.*   [M. T. 1927]

We have already seen that but for the small extra term the orbit would be an ellipse

$$u = \frac{\mu}{h^2}\{1 + e\cos(\theta - \varpi)\} \quad\ldots\ldots\ldots\ldots\ldots\ldots(1).$$

In this case the differential equation is

$$\frac{d^2u}{d\theta^2} + u = \frac{\mu}{h^2} + \frac{\alpha u}{h^2} \quad\ldots\ldots\ldots\ldots\ldots\ldots\ldots\ldots(2).$$

From this point we might proceed as in the last article to regard (1) as a first approximation to the solution of (2) and substitute the value of u given by (1) in the small term on the right of (2) and follow closely the procedure of the last article. But inasmuch as (2) can be solved in explicit terms we may proceed directly to the solution. Thus (2) may be written

$$\frac{d^2u}{d\theta^2} + \left(1 - \frac{\alpha}{h^2}\right)\left(u - \frac{\mu}{h^2 - \alpha}\right) = 0,$$

and, regarding $u - \dfrac{\mu}{h^2 - \alpha}$ as the dependent variable, the solution of this differential equation is by (16) of **1·7**

$$u - \frac{\mu}{h^2 - \alpha} = A\cos\left\{\sqrt{\left(1 - \frac{\alpha}{h^2}\right)}\,\theta - \varpi\right\} \quad\ldots\ldots\ldots\ldots(3),$$

where A, ϖ are constants of integration.

Neglecting α^2 this may be written

$$u = \frac{\mu}{h^2} + \frac{\alpha\mu}{h^4} + A\cos\left(\theta - \frac{\alpha\theta}{2h^2} - \varpi\right) \quad\ldots\ldots\ldots\ldots(4),$$

or

$$u = \frac{\mu}{h^2} + \frac{\alpha\mu}{h^4} + A\cos(\theta - \varpi - \delta\varpi) \quad\ldots\ldots\ldots\ldots\ldots(5),$$

provided that $\dfrac{\delta\varpi}{\theta} = \dfrac{\alpha}{2h^2}$.

The form of (5) shews that at any instant the orbit may be regarded as an ellipse whose apse line is advancing at the rate of $\alpha\pi/h^2$ in a revolution.

12·75. Principles of Energy and Momentum applied to Central Orbits.

Problems on central orbits can also be solved by applying the principles of energy and momentum. Thus, if f denote the force per unit mass towards the origin, the work done in a small displacement ds is

$$-f\frac{dr}{ds}\,ds, \quad\text{or}\quad -f\,dr;$$

and since $\dot{r}$ and $r\dot{\theta}$ are the components of velocity we have

$$\tfrac{1}{2}(\dot{r}^2 + r^2\dot{\theta}^2) = -\int f\,dr + C \quad\ldots\ldots\ldots\ldots(1),$$

where C is a constant.

Also the moment about O of the momentum is constant, therefore

$$r^2\dot{\theta} = h \quad\ldots\ldots\ldots\ldots\ldots\ldots\ldots(2).$$

Equations (1) and (2) are sufficient for the determination of the orbit, if f is a given function of r.

12·751. Example. *A particle moves in an orbit under a central acceleration μ/r^2 along the radius vector r; obtain the equations of energy and angular momentum.*

If the particle is projected with velocity u at right angles to the radius, at distance c from the origin, prove that

$$\left(\frac{dr}{dt}\right)^2 = \left\{\frac{2\mu}{c} - u^2\left(1+\frac{c}{r}\right)\right\}\left(\frac{c}{r}-1\right). \qquad \text{[M. T. 1913]}$$

The equation of energy is

$$\tfrac{1}{2}(\dot{r}^2 + r^2\dot{\theta}^2) = -\int \frac{\mu}{r^2}\,dr + C,$$

or $\qquad\qquad \tfrac{1}{2}(\dot{r}^2 + r^2\dot{\theta}^2) = \frac{\mu}{r} + C \quad\ldots\ldots\ldots\ldots\ldots\ldots(1),$

and the equation of angular momentum is

$$r^2\dot{\theta} = h \quad\ldots\ldots\ldots\ldots\ldots\ldots\ldots(2).$$

Initially $\qquad\qquad r=c, \quad \dot{r}=0, \quad \text{and} \quad r\dot{\theta}=u.$

Substituting in (1) we find $\quad \tfrac{1}{2}u^2 = \frac{\mu}{c} + C,$

and in (2) $\qquad\qquad\qquad\qquad cu = h\,;$

therefore $\qquad\qquad \tfrac{1}{2}(\dot{r}^2 + r^2\dot{\theta}^2) = \frac{\mu}{r} + \tfrac{1}{2}u^2 - \frac{\mu}{c},$

and $\qquad\qquad\qquad\qquad\qquad r^2\dot{\theta} = cu.$

Eliminating $\dot{\theta}$ we find that

$$\dot{r}^2 = 2\mu\left(\frac{1}{r} - \frac{1}{c}\right) + u^2 - \frac{c^2 u^2}{r^2}$$

$$= \left\{\frac{2\mu}{c} - u^2\left(1+\frac{c}{r}\right)\right\}\left(\frac{c}{r}-1\right).$$

12·8. Repulsive Forces. The motion of a particle under the action of a central repulsive force can be discussed in like manner.

For example, when the repulsive force is inversely as the square of the distance, we may write

$$\frac{h^2}{p^3}\frac{dp}{dr} = -\frac{\mu}{r^2},$$

so that $\dfrac{h^2}{p^2} = C - \dfrac{2\mu}{r}$ ...(1).

Now in **12·52** we see that for a hyperbola

$$\frac{b^2}{p^2} = \frac{r'}{r},$$

so that for points on the branch remote from the focus S we have

$$\frac{b^2}{p^2} = \frac{r - 2a}{r} = 1 - \frac{2a}{r} \quad\ldots\ldots\ldots\ldots\ldots(2),$$

and a comparison of equations (1), (2) shews that

$$C = \mu/a, \quad \text{and} \quad h^2 = \mu b^2/a.$$

Hence the particle would describe the farther branch of the hyperbola under a repulsive force μ/r^2, and from (1) the velocity would be given by

$$v^2 = \frac{\mu}{a} - \frac{2\mu}{r}.$$

12·9. Motion of Two Particles under their Mutual Attraction. If two particles are moving under the action of no forces but their mutual attraction, there is no external force acting upon the system as a whole and therefore the centre of gravity G is either at rest or moves uniformly in a straight line, **9·3**.

If the velocities of the particles at any instant are known the velocity of the centre of gravity can be calculated. If the velocity of the centre of gravity reversed in direction is compounded with the velocity of each particle, the centre of gravity is reduced to rest and the velocities of the particles become their velocities relative to the centre of gravity. The acceleration of each particle can then be expressed in terms of its distance from the centre of gravity and thus everything necessary for the determination of the orbits relative to the centre of gravity is known.

12·91. Law of Inverse Square. Thus if m, m' are the masses of the particles P and Q and G is their centre of gravity, and $PG = r$, $QG = r'$ and $\gamma mm'/(PQ)^2$ is the force between the particles, we have

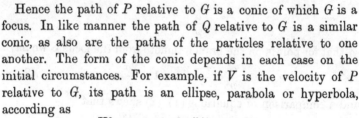

$$\frac{r}{m'} = \frac{r'}{m} = \frac{PQ}{m+m'};$$

and the acceleration of P is $\dfrac{\gamma m'}{PQ^2}$ or $\dfrac{\gamma m'^3}{(m+m')^2\, r^2}$ towards G.

Hence the path of P relative to G is a conic of which G is a focus. In like manner the path of Q relative to G is a similar conic, as also are the paths of the particles relative to one another. The form of the conic depends in each case on the initial circumstances. For example, if V is the velocity of P relative to G, its path is an ellipse, parabola or hyperbola, according as

$$V^2 <, = \text{ or } > 2m'^3/(m+m'^2)\, r.$$

If on the other hand we wish to find directly the path of P relative to Q, we can do so by first giving to both particles the reversed velocity and acceleration of Q, so that Q is thereby reduced to rest. Then the path determined for P is its path relative to Q.

12·92. Example. *Two gravitating particles of masses M, m encounter one another after approach from a great distance. If at a great distance the velocity of one relative to the other is V along a line distant p from this other particle, and ϕ is the angle through which the relative velocity is turned by the complete encounter, prove that*

$$\tan \tfrac{1}{2}\phi = \gamma\,(M+m)/\,V^2 p.$$

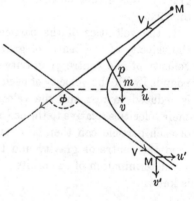

Shew further that if the two particles have equal masses and one is initially at rest the final velocities are

$$V \cos \tfrac{1}{2}\phi \quad \text{and} \quad V \sin \tfrac{1}{2}\phi.$$

The force between the particles is $\gamma Mm/r^2$, so that the accelerations of M and m are $\gamma m/r^2$ and $\gamma M/r^2$ towards one another; and the acceleration of M relative to m is $\gamma\,(M+m)/r^2$. And since at a great distance

$$V^2 > 2\gamma\,(M+m)/r,$$

if r is taken large enough, therefore the orbit of M relative to m is a hyperbola, and M is initially moving in the direction of an asymptote with velocity V and the asymptote is at a distance p from the focus. Hence p is the semi-conjugate axis of the hyperbola.

Again if v is the relative velocity at distance r

$$v^2 = \gamma \left(M+m\right)\left(\frac{2}{r}+\frac{1}{a}\right) \dots\dots\dots\dots\dots\dots(1),$$

and by substituting the values $r=\infty$, $v=V$ we get

$$V^2 = \frac{\gamma\left(M+m\right)}{a}.$$

The angle ϕ is the angle between the asymptotes indicated in the figure, because after the encounter M will be at a great distance in the direction of the second asymptote.

Hence $\tan \tfrac{1}{2}\phi = \dfrac{a}{p} = \dfrac{\gamma\left(M+m\right)}{V^2 p}.$

Again let m be initially at rest and let the final velocities of m and M parallel to the axes of the hyperbola be u, v and u', v' as indicated in the figure. The magnitude of the final relative velocity is given by (1) when $r \to \infty$ and is therefore V. Hence $u'-u= V\sin\tfrac{1}{2}\phi$, $v'-v= V\cos\tfrac{1}{2}\phi$. But the final momentum must be the same as the initial momentum, and, since the masses are equal, this gives $u'+u= -V\sin\tfrac{1}{2}\phi$ and $v'+v= V\cos\tfrac{1}{2}\phi$. From these equations we deduce $u'=0$, $u= -V\sin\tfrac{1}{2}\phi$, $v'= V\cos\tfrac{1}{2}\phi$, $v=0$, so that the final velocities of M and m are $V\cos\tfrac{1}{2}\phi$ and $V\sin\tfrac{1}{2}\phi$.

EXAMPLES

1. If the angular velocity of a particle about a point in its plane of motion be constant, prove that the transverse component of its acceleration is proportional to the radial component of its velocity.

2. The velocities of a particle along and perpendicular to a radius vector from a fixed origin are λr^2 and $\mu\theta^2$: find the polar equation of the path of the particle and also the component accelerations in terms of r and θ.
[S. 1926]

3. A steamer moving with constant speed, v, relative to the water passes round a lightship anchored in a tideway, keeping the lightship always dead abeam. Shew that the path of the steamer is an ellipse whose minor axis is in the direction of the tidal current and whose eccentricity is u/v. (u is the speed of the tide and we assume $u < v$.)   [S. 1924]

4. A smooth straight wire rotates in a plane with constant angular velocity about one end. Shew that a particle which is free to slip along the wire may describe an equiangular spiral.   [S. 1921]

5. A particle on a smooth table is attached to a string passing through a small hole in the table and carries an equal particle hanging vertically. The former particle is projected along the table at right angles to the string with velocity $\sqrt{2gh}$ when at a distance a from the hole. If r is the distance from the hole at time t, prove the results

(i) $2\left(\dfrac{dr}{dt}\right)^2 = 2gh\left(1 - \dfrac{a^2}{r^2}\right) + 2g\,(a-r)$,

(ii) the lower particle will be pulled up to the hole if the total length of string is less than $a + \frac{1}{2}h + \sqrt{ah + \frac{1}{4}h^2}$,

(iii) the tension of the string is $\frac{1}{2}mg\left(1 + \dfrac{2a^2h}{r^3}\right)$, m being the mass of each particle.   [S. 1926]

6. A smooth straight tube rotates in a horizontal plane about a point in itself with uniform angular velocity ω. At time $t=0$ a particle is inside the tube, at rest relatively to the tube, and a distance a from the point of rotation. Shew that at time t the distance of the particle from the point of rotation is

$$a \cosh \omega t.$$

Find the force the tube is then exerting on the particle.   [S. 1923]

7. If two particles P, Q describe the same ellipse under the same central force to the centre C, prove that the area of the triangle CPQ is invariable.   [Coll. Exam. 1902]

8. A particle is projected from a given point P under the action of forces whose accelerating effects are μPS, $\mu PS'$ directed towards fixed points S, S'. Find the magnitude and direction of the velocity of projection in order that the orbit may be an ellipse with S, S' for foci.   [M. T. 1910]

9. Two particles are describing the same ellipse about a centre of force in the centre in opposite directions, the mass of one being double that of the other. If the particles meet and coalesce at an end of the minor axis, shew that the new orbit trisects the major axis of the old.   [M. T. 1911]

10. A particle of unit mass describes an ellipse under the action of a central force μr. Shew that the normal component of the acceleration at any instant is $ab\mu^{\frac{3}{2}}/v$, where v is the velocity at that instant and a, b are the semi-axes of the ellipse.   [M. T. 1921]

11. An elastic string has one end fixed at A, passes through a small fixed ring at B and has a heavy particle attached at the other end. The unstretched length of the string is equal to AB. The particle is projected from any point in any manner. Assuming that it will describe a plane curve, shew that the curve is in general an ellipse.   [S. 1926]

12. A particle is repelled from a centre of force O with a force μr per unit mass, where r is the distance of the particle from O. Shew that, if the particle is projected from a point P in any direction with velocity $OP\sqrt{\mu}$, its path is a rectangular hyperbola with O as centre. [S. 1925]

13. Three particles P, Q, R, each of mass m, attract one another with a force μm^2 (distance). They move on a smooth horizontal plane and, when $t=0$, are at A, B, C and are then moving with velocities equal in magnitude and direction to λBC, λCA, λAB. Shew that their centre of gravity, G, remains at rest, and that each particle describes an ellipse about G in the periodic time $T = 2\pi/\sqrt{(3\mu m)}$.

Shew that the area of each ellipse is $\frac{2}{3}\lambda ST$, where S is the area of ABC. [M. T. 1922]

14. If a particle be describing an ellipse about a centre of force in the centre, shew that the sum of the reciprocals of its angular velocities about the foci is constant.

15. Prove that the earth's velocity of approach to the sun, when the earth in its orbit is at one end of the latus rectum through the sun, is approximately $18\frac{1}{2}$ miles per minute, taking the eccentricity of the earth's orbit as $1/60$, and $93,000,000$ miles as the semi-axis major of the earth's orbit. [M. T. 1909]

16. A body moves under a central force varying inversely as the square of the distance, the accelerating effect of the force at a distance of one foot being 32 f.s.s. If the body is projected at right angles to the radius at a distance of ·75 foot with velocity 8 f.s., determine the major axis and eccentricity of the orbit. What must be the velocity of projection to make the orbit a parabola ? [Coll. Exam. 1914]

17. A particle is projected from a point A at right angles to SA, and is acted on by a force varying inversely as the square of the distance towards S. If the intensity of the force is unity at unit distance, SA, and the velocity of projection is $\frac{1}{2}$, prove that the eccentricity of the orbit is $\frac{3}{4}$, and find the periodic time. [Coll. Exam. 1910]

18. A particle describes an ellipse under the action of a force to a focus; if the particle on arriving at any point in the orbit is diverted by a blow in the direction of the normal so that the tangent to the new orbit at that point passes through the second focus, prove that the latus rectum of the new path is four times that of the old. [Coll. Exam. 1911]

19. If a particle is describing an ellipse of eccentricity ·5 under the action of a force to a focus and when it arrives at an apse the velocity is doubled, shew that the new orbit will be a parabola or hyperbola according as the apse is the farther or nearer one. [Coll. Exam. 1912]

20. Two particles describe in equal times the arc of a parabola bounded by the latus rectum, one under an attraction to the focus, and the other with constant acceleration g parallel to the axis. Shew that the acceleration of the first particle at the vertex of the parabola is $\frac{16}{9}g$. [M. T. 1915]

21. In an elliptic orbit described under the action of a force to a focus S, if the tangent at P meet any line through S in T, shew that the component of the velocity at P in a direction perpendicular to ST varies inversely as ST. Also, find the two points of the orbit at which the component of velocity in any direction LM and the component in the opposite direction ML have maximum values; and shew that the sum of the two maximum values is the same for all directions of LM. [M. T. 1915]

22. A particle is projected from infinity with velocity V so as to pass a fixed point at a distance c if undisturbed. If it is attracted to the fixed point with acceleration μu^2, where u is the reciprocal of the radius vector, shew that the equation of the orbit is

$$u = \frac{\mu}{V^2 c^2} + \frac{\cos\theta}{c}\sqrt{\left(1 + \frac{\mu^2}{V^4 c^2}\right)},$$

θ being measured from the apse. [M. T. 1924]

23. Shew that, in elliptic motion about a focus under attraction μr^{-2}, the radial velocity is given by the equation

$$r^2\left(\frac{dr}{dt}\right)^2 = \frac{\mu}{a}\{a(1+e)-r\}\{r-a(1-e)\}.$$ [M. T. 1916]

24. Shew that, if a particle describes an ellipse under a force to a focus, the velocity at the mean distance from the centre of force is a mean proportional between the velocities at the ends of any diameter.

25. A planet of small mass is describing an elliptical orbit round a sun of large mass. When the planet is at perihelion, the mass of the sun is suddenly doubled. Shew that the planet will continue to describe an elliptical orbit, but that its velocity at perihelion is to its former velocity at aphelion as twice the major axis of the former orbit is to the major axis of the new one. [M. T. 1926]

26. A particle acted on by a central attractive force μu^3 is projected with a velocity $\frac{1}{a}\sqrt{\mu}$ at an angle of $\frac{1}{4}\pi$ with its initial distance a from the centre of force: shew that its orbit is the equiangular spiral $r = ae^{-\theta}$. [Coll. Exam. 1909]

27. If a particle is projected from an apse at distance a with the velocity from infinity under the action of a central force μr^{-2n-3}, prove that the path is $r^n = a^n \cos n\theta$. [Coll. Exam. 1910]

28. Find the velocity necessary for the description of a circular orbit of radius a under a central force $2\mu a^2 u^5 - \mu u^3$. Shew that the orbit is unstable and that if a slight disturbance takes place inwards the path may be represented by $r = a \tanh \theta$.　　　　　　　　[Coll. Exam. 1914]

29. A particle of mass m moves under a central force

$$m\mu \{3au^4 - 2(a^2 - b^2)u^5\},$$

a being $> b$, and is projected from an apse at distance $a+b$, with velocity $\sqrt{\mu}/(a+b)$; shew that its orbit is

$$r = a + b \cos \theta.$$

30. Prove that, if the law of force is $3\mu u^3 + 2\mu a^2 u^5$ and a particle is projected in a direction making an angle $\cot^{-1} 2$ with the initial distance a, and with a velocity equal to that in a circle at the same distance, the orbit is

$$au = \tan(\theta + \tfrac{1}{4}\pi).$$

31. The law of force is μu^5 and a particle is projected from an apse at distance a. Find the orbit (i) when the velocity of projection is $\sqrt{\mu}/a^2 \sqrt{2}$, (ii) when it is $\sqrt{\mu}/a^2$.

32. If the law of force be $2\mu(u^3 - a^2 u^5)$ and the particle be projected from an apse at distance a with velocity $\sqrt{\mu}/a$, shew that it will be at a distance r from the centre after a time

$$\frac{1}{2\sqrt{\mu}} \{r\sqrt{(r^2 - a^2)} + a^2 \cosh^{-1} r/a\}.$$

33. From the fact that the velocity in the hodograph of the motion of a particle represents the acceleration in the path, deduce that, if $r = f(\theta)$ be the hodograph of a projectile with the initial line horizontal and θ measured positively downwards, the retardation due to the resistance of the air is $g\left(\sin\theta - \dfrac{\cos\theta}{r}\dfrac{dr}{d\theta}\right)$, and the horizontal range is $\dfrac{1}{g}\int r^2 d\theta$.
　　　　　　　　[M. T. 1919]

34. A point describes a circle of radius a so that its hodograph is a second circle of radius b. If the pole of the hodograph be at distance c from its centre, where c/b is small, shew that the time of a complete revolution is approximately

$$2\pi a (1 + \tfrac{3}{4} c^2/b^2)/b.$$　　　　　　　　[S. 1915]

35. Prove that the hodograph of the motion of a small ring, moving along a circular wire under the action of no force, other than the reaction of the wire, is a circle or an equiangular spiral, according as the wire is smooth or not.　　　　　　　　[Coll. Exam. 1910]

36. A particle of mass m is connected to a slightly extensible string of modulus λ, the other end of which is fixed to a point in a smooth horizontal table. The particle lies on the table and is projected with velocity v at right angles to the string which is initially just taut. Shew that the maximum extension of the string is approximately $2mv^2/\lambda$.   [S. 1923]

37. Two particles of masses m, m' are moving in the plane of xy, under an attraction R along the radius r joining the particles. Shew that the centre of gravity moves with constant velocity in a straight line; and that if x, y are the rectangular coordinates of either particle with respect to the other, then

$$\frac{mm'}{m+m'}\frac{d^2x}{dt^2} = -R\frac{x}{r}, \quad \frac{mm'}{m+m'}\frac{d^2y}{dt^2} = -R\frac{y}{r}.$$

If the relative orbit is a circle of radius r, described in a period T, prove that

$$R = \frac{mm'}{m+m'}\frac{4\pi^2 r}{T^2}.$$

Assuming Newton's law of attraction, and that the moon describes a circle of radius r, relative to the earth, establish the equation

$$1 + \frac{M}{E} = \frac{4r^3}{a^2 l N^2},$$

where a is the earth's radius, l the length of the seconds pendulum, M and E are the masses of the moon and earth respectively, and N is the number of seconds in the moon's period.   [M. T. 1914]

38. Two gravitating particles of masses m and M, starting from rest at an infinite distance apart, are allowed to fall freely towards one another. Shew that when their distance apart is a, their relative velocity of approach is $\sqrt{\{2\gamma(M+m)/a\}}$, γ being the constant of gravitation.

Impulses of magnitude I are applied to the particles in opposite directions perpendicular to their line of motion at the instant when their distance apart is a. Shew that p, their distance of closest approach during the subsequent motion, is the positive root of a certain quadratic equation, and that if I is small, p is given approximately by

$$\frac{p}{a} = \frac{1}{2}\frac{I^2 a}{\gamma}\frac{M+m}{M^2 m^2}.$$ [M. T. 1923]

39. A weight can slide along the spoke of a horizontal wheel but is connected to the centre of the wheel by means of a spring. When the wheel is fixed the period of a small oscillation of the weight is $2\pi/n$: shew that when the wheel is made to rotate with constant angular velocity ω the period of oscillation is

$$2\pi/\sqrt{(n^2 - \omega^2)}.$$

If the wheel is a light frame whose mass may be neglected, and is started to rotate freely with angular velocity Ω, shew that if $\Omega = 6n/5\sqrt{11}$ the greatest stretch of the spring is $20\,^\circ/_\circ$ of its original length.

[M. T. 1917]

ANSWERS

2. $\dfrac{1}{2\lambda r^2} = \dfrac{1}{\mu\theta} + C$; $2\lambda^2 r^3 - \mu^2\theta^4/r$; $\lambda\mu r\theta^2 + 2\mu^2\theta^3/r$.   6. $2ma\omega^2\sinh\omega t$.

8. $\sqrt{(2\mu SP \cdot S'P)}$.   16. 3 ft.; 0·5; $16/\sqrt{3}$ f.s.   17. $16\pi/7\sqrt{7}$ secs.

21. The points are the ends of the chord through S perpendicular to LM.

28. $\sqrt{\mu}/a$.   31. (i) $r = a\cos\theta$; (ii) $\tanh\theta/\sqrt{2} = r/a$ or a/r both having a common asymptotic circle $r = a$.

Chapter XIII

MOMENTS OF INERTIA

13·1. If the mass of every element of a body or particle of a system be multiplied by the square of its distance from an axis, the sum of the products is called the **Moment of Inertia** of the body or system about that axis. Thus if m denotes the mass of an element or particle and r denotes its distance from the axis, the moment of inertia is Σmr^2.

In like manner we may define the moment of inertia of a system with respect to a point or plane as Σmr^2, where r denotes distance from the point or plane.

If M denotes the whole mass and k be a line of such length that Mk^2 is the moment of inertia about an axis, then k is called the **radius of gyration** of the system about that axis.

When rectangular coordinate axes are used, the moments of inertia of a body about the axes are denoted by

$$A = \Sigma m (y^2 + z^2), \quad B = \Sigma m (z^2 + x^2), \quad C = \Sigma m (x^2 + y^2);$$

and the sums represented by

$$D = \Sigma myz, \quad E = \Sigma mzx, \quad F = \Sigma mxy$$

are called the **products of inertia** of the body with regard to the axes yz, zx, xy.

13·2. Theorem of Parallel Axes. *The moment of inertia of a body about any axis is equal to its moment of inertia about a parallel axis through its centre of gravity, together with the product of the whole mass and the square of the distance between the axes.*

Let the given axis be taken as axis of z. Let (x, y, z) be the coordinates of a particle of mass m, $(\bar{x}, \bar{y}, \bar{z})$ the coordinates of the centre of gravity G, and let $x = \bar{x} + x'$, $y = \bar{y} + y'$, $z = \bar{z} + z'$.

The moment of inertia about Oz

$$= \Sigma m (x^2 + y^2) = \Sigma m \{(\bar{x} + x')^2 + (\bar{y} + y')^2\}$$
$$= \Sigma m (x'^2 + y'^2) + (\bar{x}^2 + \bar{y}^2) \Sigma m + 2\bar{x} \Sigma m x' + 2\bar{y} \Sigma m y'.$$

But $\Sigma mx' = 0$ and $\Sigma my' = 0$, therefore

The moment of inertia about $Oz = \Sigma m (x'^2 + y'^2) + (\bar{x}^2 + \bar{y}^2) \Sigma m$;

and the first sum is the moment of inertia about an axis through G parallel to Oz, and the remaining terms are the product of the whole mass and the square of the distance between the axes.

13·3. Plane Lamina. *The moment of inertia of a plane lamina about an axis perpendicular to its plane is equal to the sum of the moments of inertia about any two perpendicular axes in the plane that intersect on the first axis.*

Take the plane of the lamina as the plane of xy and the perpendicular axis as Oz. Then, since $z = 0$ at all points on the lamina, we have

$$A = \Sigma my^2, \quad B = \Sigma mx^2, \quad C = \Sigma m (x^2 + y^2),$$

so that
$$C = A + B.$$

13·4. Reference Table. The calculation of a moment of inertia is generally a simple matter of integration and the following results are tabulated for convenience. In all cases Mk^2 is taken to be the moment of inertia, where M is the whole mass, and G denotes the centre of gravity. In some of the cases given the result follows from the preceding result by an application of **13·2** or **13·3**.

	Value of k^2
Rod of length $2a$,	
about a perpendicular axis through G	$\tfrac{1}{3}a^2$
about a perpendicular axis through an end	$\tfrac{4}{3}a^2$
Rectangular lamina of sides $2a$, $2b$,	
about an axis bisecting the sides $2a$	$\tfrac{1}{3}a^2$
about a perpendicular to its plane through G	$\tfrac{1}{3}(a^2+b^2)$
about a perpendicular to its plane through a corner ...	$\tfrac{4}{3}(a^2+b^2)$
Rectangular parallelopiped of edges $2a$, $2b$, $2c$,	
about an axis through its centre perpendicular to the plane containing the edges $2b$, $2c$	$\tfrac{1}{3}(b^2+c^2)$
Elliptic lamina of axes $2a$, $2b$,	
about the axis $2a$	$\tfrac{1}{4}b^2$
about the axis $2b$	$\tfrac{1}{4}a^2$
about a perpendicular to its plane through G	$\tfrac{1}{4}(a^2+b^2)$
Ellipsoid of axes $2a$, $2b$, $2c$,	
about the axis $2a$	$\tfrac{1}{5}(b^2+c^2)$

The last three results may be included in a single formula known as **Routh's rule**:

A body has three axes of symmetry: about an axis of symmetry k^2 has the value

$$\frac{\text{Sum of squares of perpendicular semi-axes}}{3, 4 \text{ or } 5},$$

where the denominator is to be 3, 4 or 5, according as the body is rectangular, elliptical or ellipsoidal.

Thus a circle is a special case of an ellipse, and if we want the radius of gyration of a circular lamina of radius a about a diameter, the perpendicular semi-axis in the plane is of length a and that perpendicular to the plane is zero, therefore $k^2 = \frac{1}{4}a^2$. But for an axis through the centre perpendicular to the plane

$$k^2 = \frac{1}{4}(a^2 + a^2) = \frac{1}{2}a^2.$$

We notice that a sphere is a particular case of an ellipsoid and that its moment of inertia about a diameter is therefore $\frac{2}{5}Ma^2$, where a is the radius.

13·41. Some of the results of the last article can be obtained by simple direct methods.

Circular disc of radius a and mass M. First find the moment of inertia about an axis through the centre perpendicular to the plane. For this purpose divide the disc into narrow concentric rings. Let r denote the radius and dr the breadth of one of these rings. Its area is $2\pi r\, dr$ and therefore its mass is $\dfrac{M}{\pi a^2} 2\pi r\, dr$ and every element of the ring is at a distance r^2 from the axis. Therefore the ring makes a contribution $\dfrac{2M}{a^2} r^3 dr$ to the moment of inertia, and the moment of inertia of the whole disc

$$= \frac{2M}{a^2} \int_0^a r^3 dr = \frac{1}{2} Ma^2.$$

Secondly, for the moment of inertia about a diameter; since the moment of inertia about every diameter is the same, therefore by **13·3** the moment of inertia about a diameter $= \frac{1}{4}Ma^2$.

Sphere of radius a and mass M. The moments of inertia about all diameters are the same, so that

$$A = B = C = \tfrac{2}{3} \Sigma m \,(x^2 + y^2 + z^2) = \tfrac{2}{3} \Sigma m r^2.$$

Divide the sphere into concentric shells. Let r denote the radius and dr the thickness of such a shell. Its volume is $4\pi r^2 dr$ and therefore its mass

$$= \frac{M}{\frac{4}{3}\pi a^3} 4\pi r^2 dr = \frac{3M}{a^3} r^2 dr.$$

Hence, for the whole sphere,

$$\Sigma m r^2 = \frac{3M}{a^3} \int_0^a r^4 dr = \tfrac{3}{5} M a^2,$$

and therefore the moment of inertia about a diameter $= \tfrac{2}{5} M a^2$.

13·42. We append two examples of integration:

Elliptic Lamina. To find the moment of inertia about the minor axis.
Let M be the mass and $2a$, $2b$ the axes.
The area may be divided into narrow strips
such as PP', in the figure, of length $2y$ and
breadth dx. The mass of such a strip is
$\dfrac{M}{\pi ab} 2y\,dx$. Hence the moment of inertia

about $Oy = \dfrac{2M}{\pi ab} \displaystyle\int_{-a}^{a} x^2 y\,dx$, where

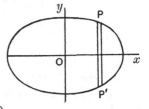

$$y = b \sqrt{\left(1 - \frac{x^2}{a^2}\right)}.$$

Therefore $Mk^2 = \dfrac{2M}{\pi a^2} \displaystyle\int_{-a}^{a} x^2 \sqrt{(a^2 - x^2)}\,dx.$

Put $x = a \sin\theta$, and we get

$$k^2 = \frac{2}{\pi} a^2 \int_{-\frac{1}{2}\pi}^{\frac{1}{2}\pi} \sin^2\theta \cos^2\theta\,d\theta$$

$$= \frac{1}{2\pi} a^2 \int_0^{\frac{1}{2}\pi} (1 - \cos 4\theta)\,d\theta$$

$$= \tfrac{1}{4} a^2.$$

It follows that the square of the radius of gyration about Ox is $\tfrac{1}{4} b^2$ and
about a perpendicular to the plane through O it is $\tfrac{1}{4}(a^2 + b^2)$.

Ellipsoid. Let M be the mass of the ellipsoid and $2a$, $2b$, $2c$ the
lengths of its axes. Taking co-
ordinate axes Ox, Oy, Oz along
the axes of the ellipsoid let us find
the moment of inertia about Ox.
Let the solid be divided by planes
parallel to yz and let $QRQ'R'$ be
the section at a distance $ON = x$
from O. The semi-axes of this
section are given by

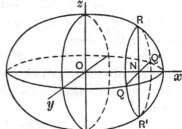

$$QN = b \sqrt{\left(1 - \frac{x^2}{a^2}\right)}$$

and $RN = c \sqrt{\left(1 - \frac{x^2}{a^2}\right)},$

so that the area of the section $=\dfrac{\pi bc}{a^2}(a^2-x^2)$, and the volume of a slice of

thickness dx is $\dfrac{\pi bc}{a^2}(a^2-x^2)\,dx$. But the whole volume is $\frac{4}{3}\pi abc$, therefore

the mass of the slice at distance x from O is $\dfrac{3M}{4a^3}(a^2-x^2)\,dx$, and the square

of its radius of gyration about Ox is

$$\tfrac{1}{4}(QN^2+RN^2)=\frac{1}{4a^2}(b^2+c^2)(a^2-x^2).$$

Hence the moment of inertia of the whole ellipsoid about Ox

$$=\frac{3M}{16a^5}(b^2+c^2)\int_{-a}^{a}(a^2-x^2)^2\,dx$$

$$=\frac{3M}{16a^5}(b^2+c^2)\left[a^4x-\tfrac{2}{3}a^2x^3+\tfrac{1}{5}x^5\right]_{-a}^{a}$$

$$=\tfrac{1}{5}M(b^2+c^2).$$

13·5. Plane Lamina. Momental Ellipse. Reverting to the
case of a plane lamina we can shew
that the moment of inertia about a
line through the origin making an
angle θ with the axis of x is given
by the formula

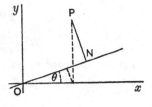

$$I=A\cos^2\theta+B\sin^2\theta-2F\sin\theta\cos\theta,$$

where A, B, F have the meanings
assigned to them in **13·1**.

For if $P(x,y)$ be the position of an element of mass m and
PN be perpendicular to the line, $PN=y\cos\theta-x\sin\theta$, and

$$I=\Sigma m\,PN^2=\Sigma m\,(y\cos\theta-x\sin\theta)^2$$

$$=\cos^2\theta\,\Sigma my^2+\sin^2\theta\,\Sigma mx^2-2\sin\theta\cos\theta\,\Sigma mxy$$

$$=A\cos^2\theta+B\sin^2\theta-2F\sin\theta\cos\theta.$$

If along any line through O we measure off a length $OQ=r$,
such that the moment of inertia of the lamina about OQ is
inversely proportional to OQ^2, the locus of the point Q is an
ellipse called a **momental ellipse** for the lamina at the point O.

To prove this theorem we have only to put $I=M\epsilon^4/r^2$, and
therefore

$$A\cos^2\theta+B\sin^2\theta-2F\sin\theta\cos\theta=M\epsilon^4/r^2,$$

or, if x, y are the rectangular coordinates of Q,

$$Ax^2+By^2-2Fxy=M\epsilon^4,$$

which represents a central conic; and, since every radius is real by construction, therefore the conic is an ellipse. M is intended to denote the mass and ϵ is a length in order to make both sides of the equation of the same dimensions.

The constant ϵ is arbitrary, so that we have any number of similar, similarly situated, momental ellipses at the same point.

13·51. Principal Axes. When rectangular axes at a point are so chosen that the products of inertia Σmyz, Σmzx and Σmxy are all zero, these axes are called the **principal axes** at the point. In this case A, B, C are called the **principal moments of inertia.**

It can be shewn that at any point there always is a set of principal axes, but we shall confine ourselves to proving this theorem for the case of a lamina.

Reverting to the figure of **13·5**, the product of inertia with regard to the axis ON and a perpendicular axis is

$$\Sigma m\,(PN.\,NO) = \Sigma m\,(y\cos\theta - x\sin\theta)(x\cos\theta + y\sin\theta)$$
$$= \Sigma mxy\,(\cos^2\theta - \sin^2\theta) + \Sigma m\,(y^2 - x^2)\sin\theta\cos\theta$$
$$= F\cos 2\theta + \tfrac{1}{2}(A - B)\sin 2\theta,$$

and this will vanish if $\theta = \tfrac{1}{2}\tan^{-1}\dfrac{2F}{B-A}$; thus determining the principal axes at O in the plane of the lamina.

The existence of the principal axes at O also follows from the fact that when the equation of a momental ellipse is referred to its axes the coefficient of xy vanishes, but this coefficient is the product of inertia with regard to those axes. Therefore the axes of a momental ellipse at O are the principal axes at O.

13·52. Geometrical Representation. Suppose that we have constructed a momental ellipse at a point O, and that a, b are its semi-axes. By hypothesis the moment of inertia about any radius OP is $M\epsilon^4/OP^2$. But if p is the length of the perpendicular from O to the tangent parallel to OP it is a property of the ellipse that $p\,.\,OP = ab$, therefore the moment

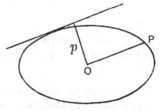

of inertia about OP is directly proportional to p^2; or, on some suitable scale of measurement, p represents the radius of gyration of the lamina about OP.

13·53. If the principal moments of inertia of a lamina at its centre of gravity are known, the moment of inertia about any other axis in the plane can be calculated. For if A, B are the principal moments and Ox, Oy the principal axes, the moment about an axis through O making an angle θ with Ox is by **13·5**

$$A \cos^2 \theta + B \sin^2 \theta,$$

and the moment about any parallel axis can be found by the theorem of parallel axes in **13·2**.

13·54. If the principal moments of inertia of a lamina are equal at any point, then a momental ellipse at that point is a circle and any pair of axes at right angles through the point are principal axes through the point. For example, at the centre of a square the moments of inertia about lines parallel to the sides of the square are clearly equal, therefore a momental ellipse is a circle and the moment of inertia about every line in the plane passing through the centre is the same.

13·6. Equimomental Bodies. Two bodies are said to be equimomental when their moments of inertia about all straight lines are equal each to each.

This will be so if the bodies have the same centre of gravity, the same mass, and the same principal axes and principal moments of inertia at the centre of gravity. Thus a straight rod of mass M is equimomental with particles of mass $\frac{1}{6}M$ at each end and a particle of mass $\frac{2}{3}M$ at the centre.

EXAMPLES

1. A parabolic area is cut off by a double ordinate at a distance h from the vertex, shew that for the moment of inertia about the tangent at the vertex $k^2 = \frac{3}{7} h^2$.

2. Shew that the square of the radius of gyration for the area of the curve $r^2 = a^2 \cos 2\theta$ about an axis through the origin perpendicular to the plane is $\frac{1}{8} \pi a^2$.

3. A right cone of height h stands on a circular base of radius a. Shew that for the moment of inertia about the axis of the cone $k^2 = \frac{3}{10}a^2$, and about a line through the vertex perpendicular to the axis $k^2 = \frac{3}{5}(h^2 + \frac{1}{4}a^2)$.

4. Shew that, for a thin hemispherical shell of radius a and mass M, the principal moments of inertia at the centre of gravity are $\frac{5}{12}Ma^2$, $\frac{5}{12}Ma^2$, $\frac{2}{3}Ma^2$; and that for a solid hemisphere the values are $\frac{83}{320}Ma^2$, $\frac{83}{320}Ma^2$, $\frac{2}{5}Ma^2$.

5. Particles of equal mass are placed at the corners of a regular polygon. Prove that the squares of the principal radii of gyration at any point O in the plane of the polygon are $a^2/2$ and $(a^2 + 2h^2)/2$, where a is the radius of the circumscribing circle of the polygon and h is the distance of O from the centre of this circle.   [Coll. Exam. 1913]

6. Prove that a uniform triangular lamina of mass M, and a system consisting of three particles each of mass $\frac{1}{3}M$, situated at the middle points of the sides and rigidly connected by light rods, have the same moment of inertia about any axis in the plane.   [M. T. 1925]

7. Shew that, if β, γ are the distances of the corners B, C of a triangle ABC from any straight line through A in the plane of the triangle, then the moment of inertia of the triangle about this line is $\frac{1}{6}M(\beta^2 + \beta\gamma + \gamma^2)$, where M denotes its mass.

8. Shew that, if α, β, γ are the distances of the corners of a triangle from any straight line in its plane, then the moment of inertia of the triangle about this line is

$$\tfrac{1}{6}M(\alpha^2 + \beta^2 + \gamma^2 + \beta\gamma + \gamma\alpha + \alpha\beta),$$

where M denotes its mass.

9. Shew that the principal axes at a corner of a rectangular lamina of sides a, b make with a side angles θ, $\frac{1}{2}\pi + \theta$, where

$$\tan 2\theta = \tfrac{3}{2}ab/(a^2 - b^2).$$

10. Shew that, in the last example, if half the rectangle be removed so as to leave a right-angled triangle of sides a, b, the value of θ for the principal axes at the right angle is given by

$$\tan 2\theta = ab/(a^2 - b^2).$$

11. A uniform rectangular plate whose sides are of lengths $2a$, $2b$ has a portion cut out in the form of a square whose centre is the centre of the rectangle and whose mass is half the mass of the plate. Shew that the axes of greatest and least moment of inertia at a corner of the rectangle make angles θ, $\frac{1}{2}\pi + \theta$, with a side, where

$$\tan 2\theta = \tfrac{6}{7}ab/(a^2 - b^2).$$ [M. T. 1928]

12. Find the moment of inertia of a uniform rod of mass m and length $2c$ about any axis for which the line of shortest distance to the given rod cuts the rod at its middle point. [Take θ for the angle between the axis and the rod.]

Shew by direct integration of the previous result that the moment of inertia of a plane uniform elliptic lamina of mass M and semi-axes a and b about an axis through its centre in the plane normal to the lamina containing the axis a and making an angle ϕ with that axis is

$$\tfrac{1}{4} M (b^2 + a^2 \sin^2 \phi).$$ [M. T. 1927]

ANSWER

12. $m (d^2 + \tfrac{1}{3} c^2 \sin^2 \theta)$, where d is the shortest distance between the rod and the axis.

Chapter XIV

MOTION OF A RIGID BODY. ENERGY AND MOMENTUM

14·1. A rigid body means a body in which the distance between each pair of particles remains invariable. The bodies with which we are familiar are all more or less elastic and capable of compression, extension or distortion under the action of external forces, and the problem of the motion of such a body is in general rendered more complicated by its deformability. The problem of the motion of bodies is greatly simplified by the hypothesis that they are rigid in the sense defined above, and in elementary dynamics we limit our considerations for the most part to the problem of the motion of one or more such bodies.

Further, we regard a rigid body as an agglomeration of particles held together by cohesive forces such that the action and reaction between any pair of particles are equal and opposite (9·7), and we use the principles established in Chapter IX for the motion of a system of particles. In the case of the rigid body however there is an additional fact to be observed, namely, that since the distance between each pair of particles is unaltered in any displacement of the body, therefore the total work done by the action and reaction between the particles is zero.

For if R denotes the mutual reaction between two particles A, B which undergo a small displacement to A', B', so that $A'B' = AB$, where $A'B'$ makes a small angle θ with AB; then, if MN is the projection of $A'B'$ on AB, the work done is

$$R \cdot AM - R \cdot BN = R(AB - MN)$$
$$= R \cdot AB(1 - \cos\theta);$$

and this is zero to the first order of small quantities, and by summation the same result holds good for a finite displacement*.

It will be convenient to summarize here the results of Chapter IX in reference to a rigid body, thus:

(i) *The rate of change of linear momentum of a rigid body in any direction is equal to the sum of the components of the external forces resolved in that direction* (**9·2**).

(ii) *The rate of change of moment of momentum of a rigid body about any fixed axis is equal to the sum of the moments of the external forces about that axis* (**9·2**).

(iii) *The increase in the kinetic energy of a rigid body in any time is equal to the work done by the external forces in that time* (**9·4**).

In **9·4** work, if any, done by the internal forces is included, but in the case of a single rigid body the internal forces do no work in any displacement and therefore in (iii) we only refer to the external forces.

The same principles also apply to the motion of a system of rigid bodies, provided that in (iii) we include the work done, if any, by the internal actions and reactions between the bodies.

The principles of the independence of translation and rotation established in **9·3** also hold good for a rigid body or system of bodies.

* A more formal proof of this theorem may be set out as follows:

Let (x, y), (x', y') be the coordinates of the particles A, B in any position of the body, R the mutual reaction and r the invariable distance of the particles, so that

$$(x - x')^2 + (y - y')^2 = r^2 \quad \ldots\ldots\ldots\ldots\ldots\ldots(1).$$

Let X, Y denote the components of R at A and consequently $-X$, $-Y$ denote the components of R at B; then

$$X = R(x' - x)/r, \quad Y = R(y' - y)/r.$$

The work done by R at A in any displacement of the body is $\int(X\,dx + Y\,dy)$ integrated along the path of A, and the work done by R at B is $-\int(X\,dx' + Y\,dy')$ integrated along the path of B. Hence the total work done is

$$-\int \frac{R}{r}\{(x' - x)(dx' - dx) + (y' - y)(dy' - dy)\};$$

and this is zero, since by differentiating (1) we get

$$(x - x')(dx - dx') + (y - y')(dy - dy') = 0.$$

14·2. In order to apply the principles of the last article, we need to find the most convenient expressions for the kinetic energy, momentum and moment of momentum of a rigid body, and we shall confine our attention for the most part to two-dimensional motions, noting however that the principles can also be applied to motion in three dimensions.

Kinetic Energy of a Rigid Body. The velocity of every point of the body is determinate provided that the angular velocity of the body is known and the linear velocity of some one point of the body.

Let us suppose that the centre of gravity G has a velocity whose components parallel to coordinate axes Ox, Oy are u, v, and let ω be the angular velocity of the body.

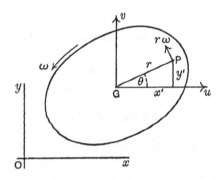

If P be the position of an element of mass m of the body whose coordinates referred to parallel axes through G are x', y' or r, θ in polar coordinates, the velocity of P relative to G is $r\omega$ perpendicular to GP. Therefore the components of velocity of P parallel to the axes are

$$u - r\omega \sin \theta \quad \text{or} \quad u - y'\omega,$$

and

$$v + r\omega \cos \theta \quad \text{or} \quad v + x'\omega.$$

Hence the kinetic energy of the body is

$$\tfrac{1}{2}\Sigma m\{(u - y'\omega)^2 + (v + x'\omega)^2\}$$
$$= \tfrac{1}{2}(u^2 + v^2)\Sigma m + \tfrac{1}{2}\omega^2\Sigma m(x'^2 + y'^2) - u\omega\Sigma my' + v\omega\Sigma mx'.$$

But Σm is the whole mass M of the body; $\Sigma mx'$ and $\Sigma my'$ are zero because G is the centre of gravity, and $\Sigma m(x'^2 + y'^2)$ is the moment of inertia of the body about an axis through G

perpendicular to the plane and may be denoted by Mk^2, where for a plane body we may speak of k as the radius of gyration about G. Hence the kinetic energy may be written

$$\tfrac{1}{2} M \left(u^2 + v^2 + k^2 \omega^2 \right) \quad \dots\dots\dots\dots(1).$$

The first two terms represent the kinetic energy of the whole mass moving with the velocity of the centre of gravity and the third term is the kinetic energy of the motion relative to the centre of gravity, in accordance with the theorem previously proved in **9·5**.

As a special case, when the body is turning about a fixed axis through O perpendicular to the plane xy, let h be the distance of G from this axis, then $h\omega$ is the velocity of G and the formula (1) becomes $\tfrac{1}{2} M (h^2 + k^2) \omega^2$, or by the theorem of parallel axes in **13·2**

$$\tfrac{1}{2} I \omega^2 \quad \dots\dots\dots\dots\dots\dots(2),$$

where I is the moment of inertia of the body about the fixed axis. This result is capable of simple independent proof, for if r denotes the distance of an element of mass m from the fixed axis, its velocity is $r\omega$, and therefore the kinetic energy of the body

$$= \tfrac{1}{2} \Sigma m r^2 \omega^2 = \tfrac{1}{2} I \omega^2.$$

14·21. Examples of Conservation of Energy. Numerous problems can be solved by the principle of energy alone; we append a few solutions.

(i) *A uniform disc is free to turn about a horizontal axis through its centre perpendicular to its plane. A particle of mass m is attached to a point in the edge of the disc. If motion starts from the position in which the radius to the particle m makes an angle α with the upward vertical, find the angular velocity when m is in its lowest position.*

Let a be the radius of the disc and Mk^2 its moment of inertia about its centre. Then, if ω is the angular velocity, the kinetic energy of the disc is $\tfrac{1}{2} Mk^2 \omega^2$, and of the particle $\tfrac{1}{2} m a^2 \omega^2$, since its velocity is $a\omega$; also, since the weight of the particle is the only force that does work, the work done in reaching the lowest position is $mga(1 + \cos \alpha)$, therefore

$$\tfrac{1}{2} (Mk^2 + m a^2) \omega^2 = mga (1 + \cos \alpha).$$

(ii) *An electric motor which gives a uniform driving torque drives a pump for which the torque required varies with the angle during each revolution according to the law $T \propto \sin^2 \theta$: the mean speed of the pump is 600 rev.*

per min. and the mean horse-power required is 8. *To limit the fluctuation of speed during each revolution, a flywheel is provided between the motor and the pump which successively stores and gives out energy. Shew that the energy thus successively stored and given out by the flywheel is approximately 70 foot-pounds.* [S. 1925]

'Torque' means couple. Let the torque required for driving the pump be given by

$$T = k \sin^2 \theta \text{ ft.-lb.},$$

where k is a constant.

The mean speed is 10 revolutions or 20π radians per second. The mean rate of working is 8 horse-power or 8×550 ft.-lb. per second. Therefore the mean couple required

$$= 8 \times 550/20\pi$$
$$= 70 \text{ ft.-lb. approximately.}$$

Now the mean value of $\sin^2 \theta$ taken over a revolution

$$= \frac{1}{2\pi} \int_0^{2\pi} \sin^2 \theta \, d\theta = \frac{1}{4\pi} \int_0^{2\pi} (1 - \cos 2\theta) \, d\theta = \tfrac{1}{2}.$$

Hence the mean value of the torque required for the pump is also $\tfrac{1}{2}k$; therefore $\tfrac{1}{2}k = 70$ ft.-lb., or $k = 140$ ft.-lb.

The torque required for driving the pump is therefore given by

$$T = 140 \sin^2 \theta \text{ ft.-lb.},$$

and varies from 0 to 140 ft.-lb. during a revolution. But the electric motor gives a uniform driving torque equal to the *mean* torque required by the pump, i.e. 70 ft.-lb., and when the pump requires a smaller torque than this the surplus energy is stored by the flywheel, to be given out again when the pump requires a torque larger than the mean. Since the torque required by the pump varies from 0 to 140 ft.-lb. and the steady torque supplied by the motor is 70 ft.-lb., it follows that 70 ft.-lb. of energy is successively stored and given out by the flywheel.

14·3. Momentum of a Rigid Body. If x, y are the coordinates of an element of mass m of the body and $\bar{x}$, $\bar{y}$ are the coordinates of the centre of gravity and M is the whole mass, we have $\Sigma mx = M\bar{x}$ and $\Sigma my = M\bar{y}$. Therefore, by differentiation, $\Sigma m\dot{x} = M\dot{\bar{x}}$ and $\Sigma m\dot{y} = M\dot{\bar{y}}$, so that if u, v denote the components of velocity of G as in **14·2**, the components of linear momentum of the whole body are Mu, Mv; i.e. the same as if the whole mass were collected into a particle at G and moved with the velocity of G.

Now consider the moment of momentum of the body about an arbitrarily chosen origin O. Reverting to the notation and figure of **14·2**, the velocities of the element m are $u - y'\omega$ and $v + x'\omega$, so that its components of momentum are $m(u - y'\omega)$ and

$m(v + x'\omega)$. Taking moments about O, the moment of momentum of the whole body

$$= \Sigma m \{x(v + x'\omega) - y(u - y'\omega)\},$$

or, since $x = \bar{x} + x'$ and $y = \bar{y} + y'$, the moment of momentum

$$= \Sigma m \{(\bar{x} + x')(v + x'\omega) - (\bar{y} + y')(u - y'\omega)\}.$$

Multiplying out and remembering that $\Sigma m x' = 0$ and $\Sigma m y' = 0$, there remains

$$\Sigma m (\bar{x} v - \bar{y} u) + \omega \Sigma m (x'^2 + y'^2)$$
$$= M (\bar{x} v - \bar{y} u) + M k^2 \omega \quad ....(1),$$

where k is the radius of gyration about G.

We observe with regard to this formula that the first term $M(\bar{x}v - \bar{y}u)$ is the moment about O of a vector localized at G and having components Mu, Mv; i.e. it is the moment about O of the components of linear momentum of the body. We also remark that the second term in formula (1), viz. $Mk^2\omega$, is independent of the choice of origin O and independent of the velocity of G. It is in fact the moment about G of the momentum in the motion of the body relative to G.

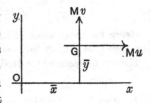

Hence the whole momentum of the body may be regarded as composed of a vector with components Mu, Mv localized at the centre of gravity, together with a couple $Mk^2\omega$ which may be called the *spin couple*; for we get the moment of momentum about any point O by taking the algebraical sum of the moments of Mu, Mv and adding on the spin couple $Mk^2\omega$.

14·31. Examples of Conservation of Momentum. Numerous problems are simple illustrations of the principles of conservation of momentum. We append the solution of a few examples.

(i) *A thin straight tube of small bore is movable about its centre on a smooth horizontal table, and it contains a uniform thin rod of the same length and mass whose centre of gravity is nearly at the middle point of the tube. Prove that, if the system be set in motion with angular velocity ω, the angular velocity of the tube as the rod leaves it is $\frac{1}{7}\omega$.*

Let m be the mass and $2a$ the length of either body. The initial moment
of momentum about O is $\frac{2}{3}ma^2\omega$.
If ω' is the angular velocity required, when the rod is leaving
the tube its centre of gravity G
has velocity $2a\omega'$, so that the momentum of the rod consists of a
linear component $2ma\omega'$ at right
angles to the rod and a spin couple

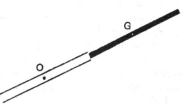

$\frac{1}{3}ma^2\omega'$, hence the moment about O of the momentum of the rod is
$4ma^2\omega' + \frac{1}{3}ma^2\omega'$, and the moment of momentum of the tube is now $\frac{1}{3}ma^2\omega'$.
And since there is no external force the total moment of momentum about
O remains unaltered, therefore

$$4ma^2\omega' + \tfrac{2}{3}ma^2\omega' = \tfrac{2}{3}ma^2\omega,$$

or $$\omega' = \tfrac{1}{7}\omega.$$

(ii) *A horizontal wheel with buckets on its circumference revolves about a
frictionless axis. Water falls into the buckets at a uniform rate of mass m
per unit of time. Treating the buckets as small compared with the wheel,
find the angular velocity of the wheel after time t, if Ω be its initial value;
and shew that if I be the moment of inertia of the wheel and buckets about
the vertical axis and r the radius of the circumference on which the buckets
are placed, the angle turned through by the wheel in time t is*

$$\frac{I\Omega}{mr^2}\log_e\left(1 + \frac{mr^2t}{I}\right).$$ [M. T. 1916]

In time t a mass mt of water falls into the buckets and the total moment
of inertia about the axis is increased from I to $I + mr^2t$. If θ be the angle
turned through in time t the angular velocity is $\dot{\theta}$, and the moment of
momentum about the axis at time t is $\dot{\theta}(I + mr^2t)$. But the system is not
acted upon by any force which has a moment about the axis, so that the
moment of momentum is constant; therefore

$$\dot{\theta}(I + mr^2t) = \Omega I.$$

Hence $$\theta = \int_0^t \frac{I\Omega\,dt}{I + mr^2t} = \frac{I\Omega}{mr^2}\left[\log_e(I + mr^2t)\right]_0^t$$
$$= \frac{I\Omega}{mr^2}\log_e\left(1 + \frac{mr^2t}{I}\right).$$

(iii) *A man of mass m stands at A on a horizontal lamina which can
rotate freely about a fixed vertical axis O. Originally both man and lamina
are at rest. The man proceeds to walk on the lamina, ultimately describes
(relative to the lamina) a closed circle having OA ($=a$) as diameter, and
returns to the point of starting on the lamina. Shew that the lamina has
moved through an angle relative to the ground given by $\pi\{1 - \sqrt{I/(I + ma^2)}\}$,
where I is the moment of inertia of the lamina about the axis.*   [M. T. 1921]

Let Ox be a fixed line, and at time t let $AOx=\theta$ and $AOP=\phi$, where
P denotes the position of the man.
The angle xOP increases at the rate
$\dot\phi-\dot\theta$, therefore the velocity of the man
at right angles to OP is $OP(\dot\phi-\dot\theta)$ and
his moment of momentum about O is
$mOP^2(\dot\phi-\dot\theta)$. The lamina has a moment
of momentum $I\dot\theta$ in the opposite sense;

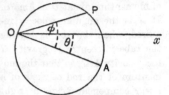

and the total moment of momentum in the same sense remains zero
throughout the motion, therefore

$$I\dot\theta-mOP^2(\dot\phi-\dot\theta)=0.$$

But $OP=a\cos\phi$, therefore

$$(I+ma^2\cos^2\phi)\,\dot\theta=ma^2\cos^2\phi\dot\phi.$$

Hence
$$d\theta=\frac{ma^2\cos^2\phi}{I+ma^2\cos^2\phi}\,d\phi$$

$$=\left(1-\frac{I}{I+ma^2\cos^2\phi}\right)d\phi.$$

Now as the man walks round the circle from A to O the angle ϕ increases
from 0 to $\frac12\pi$, and as he walks on from O to A the angle ϕ increases from
$\frac32\pi$ to 2π. Therefore the whole angle turned through by the lamina

$$=\int_0^{\frac12\pi}+\int_{\frac32\pi}^{2\pi}\left(1-\frac{I}{I+ma^2\cos^2\phi}\right)d\phi$$

$$=\pi-2I\int_0^{\frac12\pi}\frac{d\phi}{I+ma^2\cos^2\phi}.$$

To evaluate the integral put $\tan\phi=t$, and we get

$$\pi-2I\int_0^\infty\frac{dt}{I+ma^2+It^2}$$

$$=\pi-\frac{2I}{\sqrt{\{I(I+ma^2)\}}}\left[\tan^{-1}\sqrt{\left(\frac{I}{I+ma^2}\right)}\cdot t\right]_0^\infty$$

$$=\pi\{1-\sqrt{I/(I+ma^2)}\}.$$

14·4. Examples of Conservation of Energy and Momentum.
(i) *A uniform straight rod is held at an inclination α to the vertical with
its lower end in contact with a smooth horizontal plane and let go. Find
the angular velocity when the rod becomes horizontal.*

Since there is no horizontal force acting on the rod its centre of gravity G
does not acquire any horizontal velocity
and therefore remains in the same vertical
line Oy. If m is the mass, $2a$ the length,
y the height of G above the plane and θ
the inclination to the vertical at time t,
the kinetic energy is $\frac12m(\dot y^2+k^2\dot\theta^2)$, where
$k^2=\frac13a^2$ and $y=a\cos\theta$; therefore the kinetic
energy is $\frac12ma^2(\sin^2\theta+\frac13)\dot\theta^2$; and the work

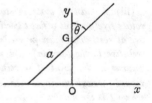

done as G descends from a height $a\cos\alpha$ to $a\cos\theta$ is $mga\,(\cos\alpha-\cos\theta)$; therefore

$$\tfrac{1}{2}ma^2\,(\sin^2\theta+\tfrac{1}{3})\,\dot\theta^2 = mga\,(\cos\alpha-\cos\theta)$$

gives the angular velocity in any position, and for $\theta=\tfrac{1}{2}\pi$

$$\dot\theta=\surd(3g\cos\alpha/2a).$$

(ii) *A sphere of radius a, and of radius of gyration k about any axis through its centre, rolls with linear velocity v on a horizontal plane, the direction of motion being perpendicular to a vertical face of a fixed rectangular block of height h, where $h<a$. The sphere strikes the block; the sphere and the block are perfectly rough and perfectly inelastic. Shew that the sphere will surmount the block if*

$$(a^2-ah+k^2)^2\,v^2 > 2gha^2\,(a^2+k^2).\qquad\text{[M. T. 1924]}$$

Since the sphere rolls with linear velocity v, therefore its angular velocity is v/a just before striking the block. Let ω be the angular velocity immediately after striking the block. If O is the point of contact of the sphere with the edge of the block the sphere begins to turn about O and when it begins to rise the velocity of its centre G is $a\omega$ at right angles to GO, and the

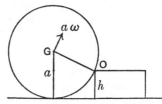

moment of momentum about O is therefore $m\,(a^2+k^2)\,\omega$, immediately after the impact, where m is the mass. Now there is an impulsive reaction on the sphere at O, but it has no moment about O, therefore the moment of momentum about O is unchanged by the impact. But before the impact there was a linear momentum mv and a spin couple mk^2v/a, therefore

$$m\,(a^2+k^2)\,\omega = mv\,(a-h)+mk^2v/a.$$

This equation determines the angular velocity ω with which the sphere begins to rise. Now in order to surmount the block the kinetic energy of the sphere when beginning to rise must be more than the work that would have to be done in lifting the sphere through a height h; otherwise the kinetic energy would be destroyed by the time the sphere had risen to the required height. But since the velocity of G is $a\omega$ when starting to rise, therefore the kinetic energy of the sphere is $\tfrac{1}{2}m\,(a^2+k^2)\,\omega^2$, and the required condition is $\tfrac{1}{2}m\,(a^2+k^2)\,\omega^2 > mgh$, and, on substituting for ω, we get

$$(a^2-ah+k^2)^2\,v^2 > 2gha^2\,(a^2+k^2).$$

EXAMPLES

1. A wheel whose moment of inertia about its axis is 200 lb.-ft.^{2} makes 10 revolutions per second. What is its kinetic energy (in ft.-lb.), and what constant couple would reduce it to rest in one revolution ?

2. A flywheel of a pressing machine has 150,000 ft.-lb. of kinetic energy stored in it when its speed is 250 revolutions per minute. What energy does it part with during a reduction in speed to 200 revolutions per minute ? If 82 °/$_0$ of this energy given out is imparted to the pressing rod during a stroke of 2 inches, what is the average force exerted by the rod ?   [M. T. 1917]

3. AB and CD are two rods of lengths $2a$ and b respectively, the mass of each rod being m per unit length. The rods are rigidly joined together at right angles at C, the middle point of AB, and the system is free to rotate in a vertical plane about D. If the system is held with AB vertical and then let go, calculate the angular velocity when AB is horizontal.
                                                          [M. T. 1912]

4. A wheel and axle has moment of inertia I about its axis. A mass m is suspended from the axle, which is of radius r, by a fine thread wrapped round and fastened to the axle. Rotation is opposed by a constant frictional couple G. Find the angular velocity acquired as the mass m descends a distance h.

5. A uniform sphere rotating about a diameter contracts by cooling. Shew that when its radius is reduced to $\frac{1}{n}$-th of its former value the kinetic energy is multiplied by n^2.

6. A wheel 30 inches in diameter, which can rotate in a vertical plane about a horizontal axis through its centre O, carries a mass of $\frac{1}{4}$ lb. concentrated at a point P on its rim. The wheel is held with OP inclined at 30° above the horizontal and then released. Owing to a friction couple of constant magnitude L at the bearing, the first swing carries OP to a position only 45° beyond the vertical. Determine the value of L; and prove that in the next swing OP will come to rest before reaching the vertical.   [M. T. 1913]

7. A flywheel of moment of inertia I is rotating with angular velocity Ω about a vertical axis. The flywheel contains a pocket at a distance a from the axis into which is dropped a sphere of mass M, moment of inertia i and spin ω about a vertical axis, without horizontal motion. Find the angular velocity of the system after the sphere has come to relative rest in the pocket.   [S. 1927]

8. A uniform rod of length a and mass m is describing circles on a smooth horizontal table about one end which is fixed. If the rod strikes a particle of mass m' at distance b from the fixed end, and this particle adheres to the rod, find the ratio in which the angular velocity is changed.

[Coll. Exam. 1908]

9. A cube of side $2a$ slides down a smooth plane inclined at an angle $2\tan^{-1}\frac{1}{5}$ to the horizontal, and meets a fixed horizontal bar placed perpendicular to the plane of motion and at a perpendicular distance $\frac{1}{4}a$ from the plane. Shew that, if the cube is to have sufficient velocity to surmount the obstacle when it reaches it, it must be allowed first to slide down the plane through a distance $\frac{107}{50}a$. The obstacle may be taken to be inelastic and so rough that the cube does not slip on it. [M. T. 1918]

10. A uniform cube of edge $2a$ is placed in unstable equilibrium with one edge in contact with a horizontal plane and allowed to fall. Shew that, if ω is the angular velocity when a face of the cube comes into contact with the plane, then

$$\omega^2 = \tfrac{6}{5}(\sqrt{2}-1)\,g/a \quad \text{or} \quad \tfrac{3}{4}(\sqrt{2}-1)\,g/a,$$

according as the plane is smooth or sufficiently rough to prevent sliding.

ANSWERS

1. 12,500 ft.-lb. ; $6250/\pi$ ft.-lb.   2. 54,000 ft.-lb. ; $118\tfrac{17}{28}$ tons.

3. $\{3gb\,(4a+b)/(2a^3+6ab^2+b^3)\}^{\frac{1}{2}}$.   4. $\{2h\,(mgr-G)/r\,(I+mr^2)\}^{\frac{1}{2}}$.

6. $15\,(1+\sqrt{2})/88\pi$ ft.-lb.   7. $(I\Omega+i\omega)/(I+i+Ma^2)$.   8. $ma^2+3m'b^2 : ma^2$.

Chapter XV

EQUATIONS OF MOTION OF A RIGID BODY

151. Using the expressions found in the last chapter for the momentum of a body, we obtain the formal equations of motion of the body by equating the rates of change of momentum to the external forces and equating the moment of the rate of change of momentum about any axis to the moment of the external forces; thus

$M\dot{u}$ = sum of the external forces resolved parallel to Ox,

$M\dot{v}$ =   „   „   „   „   „ Oy;

and an equation of moments, viz. algebraical sum of moments about O of $M\dot{u}$ and $M\dot{v}$ plus the couple $Mk^2\dot{\omega}$ = algebraical sum of moments about O of the external forces, where O is *any* point in the plane.

If we take O to be the point through which the centre of gravity G is passing at the instant considered, then since $M\dot{u}$ and $M\dot{v}$ are localized at G they have no moment about O and the last equation reduces to $Mk^2\dot{\omega}$ = sum of moments about G of the external forces.

15·2. Applications of the Equations of Motion. (i) *Pulley with inertia.* Let two particles of masses m, m' be connected by a fine string passing over a pulley of mass M, radius a and moment of inertia about its axis Mk^2. Suppose that the pulley is free to turn about its axis without friction and that the string does not slip on the pulley. Let the tensions in the two straight portions of the string be T, T'.

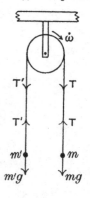

If the angular acceleration of the pulley is $\dot{\omega}$ in the sense indicated in the figure, the linear acceleration of the string and therefore also of the particles is $a\dot{\omega}$. Hence by resolving vertically for the particles we get

$$ma\dot{\omega} = mg - T \quad \text{and} \quad m'a\dot{\omega} = T' - m'g.$$

And by taking moments about its axis for the pulley we get

$$Mk^2\dot{\omega} = aT - aT',$$

and if we eliminate the tensions we get

$$(ma^2 + m'a^2 + Mk^2)\,\dot{\omega} = (m - m')\,ag,$$

shewing that the acceleration is constant.

Alternatively this result might be obtained from the principle of energy. For if we suppose that in descending a distance s the particle m acquires a velocity v, then the angular velocity acquired by the pulley is v/a, and the whole kinetic energy is

$$\tfrac{1}{2}(m+m')v^2+\tfrac{1}{2}Mk^2v^2/a^2,$$

·but the work done as m descends and m' ascends a distance s is $(m-m')gs$; therefore

$$\tfrac{1}{2}(m+m'+Mk^2/a^2)v^2=(m-m')gs,$$

and differentiating with regard to s gives

$$(m+m'+Mk^2/a^2)v\,dv/ds=(m-m')g;$$

so that the acceleration $v\,dv/ds$ has the same value as was obtained previously for $a\dot\omega$.

(ii) *A circular hoop, a disc or a sphere rolls down an inclined plane.*
Let a be the radius of the body, m the mass, mk^2 the moment of inertia, F the friction, R the normal reaction and α the inclination of the plane to the horizontal.

Let u be the velocity of the centre G and ω the angular velocity.

The first of the diagrams shews the external forces F, R and the weight mg, the second shews the rates of change of momentum, and these are two equivalent systems.

Resolving along and at right angles to the plane we get

$$m\dot u=mg\sin\alpha-F\ldots\ldots(1),$$

and

$$0=mg\cos\alpha-R\ldots\ldots(2).$$

Taking moments about G gives

$$mk^2\dot\omega=Fa\ldots\ldots(3).$$

One other equation is required for the determination of the four unknown quantities F, R, u, ω, and it is got from the kinematical condition that the body rolls, i.e. that the velocity of the point P in contact with the plane is zero. The velocity of P relative to G is $a\omega$ up the plane and the velocity of G is u down the plane, therefore the condition that the body rolls is

$$u-a\omega=0\ldots\ldots(4).$$

From (3) and (4) we get

$$F=mk^2\dot u/a^2\ldots\ldots(5),$$

and substituting for F in (1) gives

$$\left(1+\frac{k^2}{a^2}\right)\dot u=g\sin\alpha\ldots\ldots(6),$$

so that the body rolls down the plane with constant acceleration.

For a hoop $k^2 = a^2$ and $\dot{u} = \frac{1}{2} g \sin \alpha$.

   „   disc $k^2 = \frac{1}{2} a^2$ „ $\dot{u} = \frac{2}{3} g \sin \alpha$.

   „   sphere $k^2 = \frac{2}{5} a^2$ „ $\dot{u} = \frac{5}{7} g \sin \alpha$.

We have assumed that the body rolls and found the acceleration on this hypothesis, but whether it rolls or slides depends on the relation of the coefficient of friction μ to the slope of the plane. The friction F cannot exceed limiting friction μR. From (5) and (6) we see that the friction necessary to cause rolling is

$$F = \frac{mk^2}{a^2 + k^2} g \sin \alpha ;$$

but

$$R = mg \cos \alpha,$$

therefore

$$\frac{F}{R} = \frac{k^2}{a^2 + k^2} \tan \alpha,$$

and the body will slide instead of roll if this fraction exceeds μ; and the acceleration when sliding is $g (\sin \alpha - \mu \cos \alpha)$. (See **4·55**.)

(iii) *A sphere rolls down on the surface of a fixed sphere.* Let a, b be the radii of the moving and fixed spheres. Let G be the centre, m the mass, mk^2 the moment of inertia of the moving sphere, F the friction and R the normal reaction. Let the common radius OG through the point of contact P make an angle θ with the vertical at time t. Since G describes

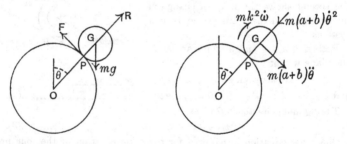

a circle of radius $a + b$, therefore it has accelerations $(a + b) \ddot{\theta}$ at right angles to OG and $(a + b) \dot{\theta}^2$ along GO. And if ω is the angular velocity of the sphere the rate of change of the spin couple is $mk^2\dot{\omega}$. The first of the two diagrams shews the external forces acting on the moving sphere, and the second shews the rates of change of momentum and these two systems are equivalent.

Resolving at right angles to and along GO we get

$$m (a + b) \ddot{\theta} = mg \sin \theta - F \quad\quad\dots\dots\dots\dots\dots\dots(1),$$

and

$$m (a + b) \dot{\theta}^2 = mg \cos \theta - R \quad\quad\dots\dots\dots\dots\dots\dots(2).$$

Taking moments about G gives

$$mk^2\dot{\omega} = Fa \quad\quad\dots\dots\dots\dots\dots\dots\dots\dots\dots(3)$$

The kinematical condition for rolling, which expresses that the point P of the moving sphere has no velocity, is

$$(a+b)\,\dot{\theta} - a\omega = 0 \quad \dotfill (4),$$

because $a\omega$ is the velocity of P relative to G and $(a+b)\,\dot{\theta}$ is the velocity of G in the opposite direction.

From (3) and (4), putting $\tfrac{2}{5}a^2$ for k^2, we get

$$F = \tfrac{2}{5}ma\dot{\omega} = \tfrac{2}{5}m\,(a+b)\,\ddot{\theta} \quad \dotfill (5)$$

and by substituting this value for F in (1)

$$\tfrac{7}{5}\,(a+b)\,\ddot{\theta} = g\sin\theta \dotfill (6).$$

Multiply by $2\dot{\theta}$ and integrate,

$$\tfrac{7}{5}\,(a+b)\,\dot{\theta}^2 = C - 2g\cos\theta.$$

Let the motion begin when $\theta = a$, i.e. $\dot{\theta} = 0$ when $\theta = a$, so that $C = 2g\cos a$, and

$$\tfrac{7}{5}\,(a+b)\,\dot{\theta}^2 = 2g\,(\cos a - \cos\theta) \quad \dotfill (7)$$

gives the velocity in any position.

The sphere will begin to slide when the friction F becomes equal to μR. But (5) and (6) give

$$F = \tfrac{2}{7}mg\sin\theta,$$

and (2) and (7) give

$$R = \tfrac{1}{7}mg\,(17\cos\theta - 10\cos a);$$

so that sliding would begin when

$$2\sin\theta = \mu\,(17\cos\theta - 10\cos a),$$

provided that the spheres are still in contact.

And the moving sphere will leave the fixed sphere in the position for which R vanishes, i.e. when $\cos\theta = \tfrac{10}{17}\cos a$.

Beginners in this branch of the subject will find it a help at first to draw separate diagrams to shew the rates of change of momentum and the equivalent system of external forces, as we have done in these last two examples.

Beginners should also be careful to remember that the angular velocity of a body moving in two dimensions is the rate of increase of the angle between a line *fixed in the body* and a line fixed in the plane. Thus the angular velocity of the moving sphere in this example is *not* $\dot{\theta}$ because GP is not a fixed line in the moving sphere.

We leave it as an exercise to the student to obtain equation (7) by the principle of energy, and we also remark that if we are not concerned with the question of the possibility of sliding we can avoid the introduction of F into the equations by taking moments about P, and thus getting the equation

$$ma\,(a+b)\,\ddot{\theta} + mk^2\dot{\omega} = mga\sin\theta,$$

which with the help of (4) is the equivalent of (6).

(iv) *A four-wheeled railway truck has a total mass M, the mass and radius of gyration of each pair of wheels and axle are m and k respectively, and the radius of each wheel is r. Prove that, if the truck is propelled along a level track by a force P, the acceleration is $P/(M+2mk^2/r^2)$; and find the horizontal force exerted on each axle by the truck.* [M. T. 1913]

[*Axle friction and wind resistance to be neglected.*]

Let u be the velocity of the truck and ω the angular velocity of the wheels. The condition that the wheels roll is, as in the last example,

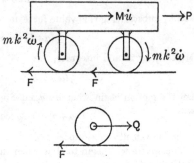

$$u - r\omega = 0 \quad \ldots\ldots(1).$$

There must be the same friction force F on each pair of wheels, since they have the same angular velocity; and by moments about the axis of a pair of wheels

$$mk^2\dot\omega = Fr \quad \ldots\ldots(2).$$

By resolving in the direction of motion for the truck and wheels

$$M\dot u = P - 2F \ldots\ldots\ldots\ldots\ldots\ldots\ldots\ldots\ldots(3).$$

But from (1) and (2)

$$F = m\dot uk^2/r^2 \ldots\ldots\ldots\ldots\ldots\ldots\ldots\ldots\ldots(4),$$

and by substituting this value of F in (3) we get

$$\dot u = P/(M + 2mk^2/r^2).$$

Again considering a pair of wheels alone, if Q be the horizontal force exerted by the truck, $m\dot u = Q - F$.

Therefore from (4)

$$Q = m\dot u\left(1 + \frac{k^2}{r^2}\right)$$
$$= \frac{m\,(r^2 + k^2)\,P}{Mr^2 + 2mk^2}.$$

(v) *The end of a thread wound round a reel is held fixed and the reel is allowed to fall so that the thread is unwound. Find the acceleration of the reel, assuming its axis to remain horizontal.*

Let m be the mass and a the radius of the reel and mk^2 its moment of inertia about its axis. Let v be the velocity of its centre and ω its angular velocity at time t.

The point A of the reel in contact with the straight thread is the instantaneous centre of rotation and has no velocity, but its velocity relative to the centre is $a\omega$, therefore

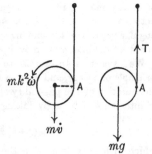

$$v - a\omega = 0 \ldots\ldots\ldots\ldots(1).$$

The rates of change of momentum are $m\dot{v}$ and the spin couple $mk^2\dot{\omega}$; and the external forces on the reel are its weight mg and the tension T of the string.

Therefore by resolving vertically we get

$$m\dot{v} = mg - T \quad \dots\dots\dots\dots\dots\dots\dots\dots\dots(2),$$

and by moments about the axis

$$mk^2\dot{\omega} = aT \quad \dots\dots\dots\dots\dots\dots\dots\dots\dots(3).$$

Hence from (1) and (3)

$$T = m\dot{v}k^2/a^2,$$

and therefore from (2)

$$\dot{v} = ga^2/(a^2 + k^2).$$

Instead of equations (2) and (3) we might take moments about the point A and thus avoid the introduction of the tension T. The equation of moments about A being

$$ma\dot{v} + mk^2\dot{\omega} = mga,$$

and this with (1) gives the same value for $\dot{v}$ as before.

15·3. Equations of Impulsive Motion.

Reverting to the equations of motion for a rigid body under the action of finite forces in **15·1**, which may be written

$$M\dot{u} = X,$$

$$M\dot{v} = Y,$$

and

$$Mk^2\dot{\omega} = L,$$

where X, Y denote the sums of the components of the external forces and L the sum of their moments about the centre of gravity, if we integrate these equations with respect to t through an interval from t_0 to t, we get

$$M(u - u_0) = \int_{t_0}^{t} X\,dt,$$

$$M(v - v_0) = \int_{t_0}^{t} Y\,dt,$$

and

$$Mk^2(\omega - \omega_0) = \int_{t_0}^{t} L\,dt,$$

where u_0, v_0, ω_0 denote the values of the velocities at the beginning of the interval. If we now suppose that we are concerned with impulsive forces as defined in **11·1**, the integrals are the measures of the components of the externally applied 'impulses' and their moments and may be denoted by P, Q and H, so that we have

$$M (u - u_0) = P,$$
$$M (v - v_0) = Q,$$

and
$$Mk^2 (\omega - \omega_0) = H.$$

These three equations merely repeat what we have already demonstrated in **11·2**, that the instantaneous change of momentum both as regards linear momentum and angular momentum is the exact equivalent of the externally applied impulses.

15·4. Examples of Impulses. *Two uniform rods AB, BC of lengths 2a, 2b are smoothly jointed at B and placed on a smooth table so that ABC is a straight line. An impulse P is applied at A at right angles to AB, determine the initial velocities.*

Let m, m' be the masses of the rods AB, BC and G, H their centres. It is clear that if the angular velocities of the rods are known and the linear velocity of any one point, then the velocities of every point of either rod

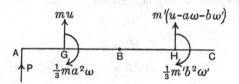

can be stated. Suppose then that u is the velocity of the point G and that ω, ω' are the angular velocities of the rods in the senses indicated in the figure.

By considering the velocities of A and B relative to G we see that the velocity of A is $u + a\omega$ and of B is $u - a\omega$, and we also see that H has a velocity $-b\omega'$ relative to B in the same sense, so that the velocity of H is $u - a\omega - b\omega'$.

Since the unknown impulse on the rod BC acts at the joint B it is convenient to take moments about B for BC, giving

$$m'b (u - a\omega - b\omega') - \tfrac{1}{3}m'b^2\omega' = 0,$$

or
$$u - a\omega - \tfrac{4}{3}b\omega' = 0 \dots\dots\dots\dots\dots\dots\dots\dots\dots(1).$$

We can again avoid introducing the unknown impulse at B by taking moments about A for the two rods together, getting

$$mau - \tfrac{1}{3}ma^2\omega + m' (2a + b) (u - a\omega - b\omega') - \tfrac{1}{3}m'b^2\omega' = 0,$$

or
$$u \{ma + m' (2a + b)\} - a\omega \{\tfrac{1}{3}ma + m' (2a + b)\} - 2b\omega'm' (a + \tfrac{2}{3}b) = 0 \dots(2).$$

Then by considering the linear momentum at right angles to the rods
we get
$$mu + m'(u - a\omega - b\omega') = P \quad ....(3).$$
Equations (1), (2) and (3) are sufficient for the determination of u, ω, ω'.

15·41. The effects of sudden prescribed changes also serve to
illustrate the application of the equations of impulsive motion :

(i) *A circular disc is revolving in its plane about its centre with angular
velocity* ω. *What is the new angular velocity if a point on the rim is
suddenly fixed?*

Let a be the radius, m the mass and ω' the new angular velocity.
Before the fixture the momen-
tum of the disc is simply the
spin couple $\frac{1}{2}ma^2\omega$ (since for a
disc $k^2 = \frac{1}{2}a^2$). After the fixture
the centre of gravity G has a
velocity $a\omega'$ at right angles to
GO, where O is the point fixed ;

therefore the momentum consists of a linear component $ma\omega'$ and a spin
couple $\frac{1}{2}ma^2\omega'$. The impulse that causes the change must act through the
point O that becomes fixed, therefore there can be no change of moment
of momentum about O ; hence we have
$$ma^2\omega' + \tfrac{1}{2}ma^2\omega' = \tfrac{1}{2}ma^2\omega, \quad \text{or} \quad \omega' = \tfrac{1}{3}\omega.$$

(ii) *A square lamina* $ABCD$ *rests on a smooth horizontal plane. If the
corner* A *is made to move with velocity* u *along the line* BA *produced, determine the initial angular velocity of the lamina.*

Let m be the mass and $2a$ a side of the square. Then, if ω is the
angular velocity, the centre of gravity G has
velocity $AG.\omega$ relative to the point A.
Therefore the velocity components of G are
$$u - AG\omega \cos \tfrac{1}{4}\pi,$$
or $u - a\omega$ parallel to BA, and
$$AG\omega \sin \tfrac{1}{4}\pi,$$
or $a\omega$ parallel to DA.

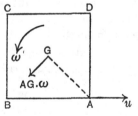

Therefore the momentum of the square
consists of linear components $m(u - a\omega)$
parallel to BA, $ma\omega$ parallel to DA and
a spin couple $mk^2\omega$, where $k^2 = \frac{2}{3}a^2$. But the
only external impulsive action is applied
at A, therefore the moment of momentum about A is zero. Hence we have

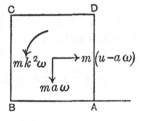

$$ma^2\omega - ma(u - a\omega) + \tfrac{2}{3}ma^2\omega = 0,$$
giving $\omega = 3u/8a.$

15·5. Motion about a Fixed Axis. When a body turns about a fixed axis the motion is completely determined by the principle that the rate of change of moment of momentum is equal to the sum of the moments of the external forces. The equation of motion takes a very simple form, for if ω be the angular velocity and r denotes the distance from the axis of an element of mass m, its velocity is $r\omega$ and its moment of momentum about the axis is $mr^2\omega$. Therefore the moment of momentum of the whole body is $\Sigma mr^2\omega$ or $I\omega$, where I is the moment of inertia of the body about the axis. Hence the equation of motion is

$$I\dot\omega = L,$$

where L is the sum of the moments about the axis of the external forces.

15·51. Compound Pendulum. Any body oscillating about a fixed horizontal axis, about which it can turn freely under the action of its weight, is called a compound pendulum.

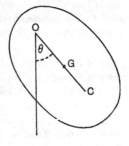

Let m be the mass of the body, G its centre of gravity at a distance h from the fixed axis. Let GO be drawn at right angles to the axis to meet it in O, and at time t let GO make an angle θ with the vertical. Also let k be the radius of gyration of the body about an axis through G parallel to the axis of rotation, then the moment of inertia about the axis of rotation is, by **13·2**, $m(k^2 + h^2)$. Hence by the last article

$$m(k^2 + h^2)\,\ddot\theta = -mgh\sin\theta,$$

or $$\frac{k^2 + h^2}{h}\,\ddot\theta = -g\sin\theta \quad\quad\dotsc\dotsc\dotsc\dotsc\dotsc(1).$$

Now in **7·6** we saw that the equation of motion of a simple pendulum of length l is

$$l\ddot\theta = -g\sin\theta;$$

therefore the length of the equivalent simple pendulum is

$$l = (k^2 + h^2)/h.$$

The period of small oscillations of the compound pendulum, i.e. oscillations of such small amplitude that we may write θ for $\sin \theta$ in (1), is

$$2\pi \sqrt{\{(k^2 + h^2)/gh\}} \quad \text{.....................(2).}$$

If on OG produced we take a point C such that

$$OC = l = h + k^2/h,$$

then C is called the **centre of oscillation**, while O is called the **centre of suspension of the pendulum**.

Since $OC = h + k^2/h$, where $h = OG$, therefore

$$OG \cdot GC = k^2 \quad \text{.......................(3),}$$

whence it follows that the centres of suspension and oscillation are interchangeable, i.e. if the body swings about a parallel axis through C the point O will become the centre of oscillation and the period of small oscillations will remain unaltered.

15·52. Pressure on the Axis. We can also determine the reaction of the axis on the body. Thus if X, Y are the components of this reaction at O at right angles to and along GO (the body is assumed to be symmetrical about the plane of

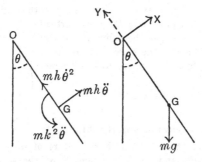

the diagram), one of the accompanying figures represents the external forces and the other the equivalent system of rates of change of momentum; and by resolving at right angles to and along GO, we get

$$mh\ddot{\theta} = X - mg \sin \theta \quad \text{..................(1),}$$

$$mh\dot{\theta}^2 = Y - mg \cos \theta \quad \text{..................(2),}$$

and by taking moments about the axis, as in **15·51**,

$$m(k^2 + h^2)\ddot{\theta} = -mgh \sin \theta \quad \text{...............(3).}$$

By multiplying the last equation by $2\dot{\theta}$ and integrating we get

$$(k^2 + h^2)\,\dot{\theta}^2 = C + 2gh\cos\theta,$$

where the constant of integration can be determined if the velocity is known in any one position; for example, if α is the amplitude of the oscillation, so that $\dot{\theta} = 0$ when $\theta = \alpha$, then $C = -2gh\cos\alpha$, and

$$(k^2 + h^2)\,\dot{\theta}^2 = 2gh\,(\cos\theta - \cos\alpha)\ \dots\dots\dots(4).$$

Substituting from (3) and (4) in (1) and (2) we get

$$X = \frac{mgk^2}{k^2 + h^2}\sin\theta,$$

and

$$Y = mg\cos\theta + \frac{2mgh^2}{k^2 + h^2}(\cos\theta - \cos\alpha).$$

15·53. Examples on the Compound Pendulum. (i) *An arc of a circle is formed of thin wire (whose density may or may not be uniform) and hangs from a point P of the arc. Shew that, if in the position of equilibrium Q is the point of the circle vertically below P, then PQ is the length of the equivalent simple pendulum when the wire oscillates about P in its own plane.* [M. T. 1916]

Let APB be the arc, G its centre of gravity, O the centre of the circle and a its radius. In equilibrium PGQ is a vertical chord of the circle. Let k be the radius of gyration of the arc about G and m the mass. The moment of inertia about O is ma^2, therefore, by the theorem of parallel axes, **13·2**,

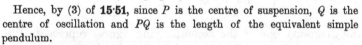

$$ma^2 = mk^2 + m.OG^2.$$

Therefore $k^2 = a^2 - OG^2$

$$= PG.GQ\ \text{by the property of the circle.}$$

Hence, by (3) of **15·51**, since P is the centre of suspension, Q is the centre of oscillation and PQ is the length of the equivalent simple pendulum.

(ii) *A pendulum consists of a thin uniform rod of mass M pivoted at its mid-point and of a regulating nut of mass m, which can be screwed to any desired position of the rod. The nut may be treated as a particle. Prove that if $M > 3m$ the period is always lengthened when the nut is raised slightly, but if $M < 3m$ the period is lengthened or shortened according to the position of the nut.* [M. T. 1923]

Let $2a$ be the length of the rod, x the distance of the nut and h the distance of the centre of gravity of the whole from the mid-point. Then

$(M+m)\,h=mx$, and the moment of inertia of the whole about the point of suspension is $\tfrac{1}{3}Ma^2+mx^2$. Therefore by formula (2) of **15·51** the period

$$=2\pi\,\sqrt{\{(\tfrac{1}{3}Ma^2+mx^2)/(M+m)\,gh\}}$$
$$=2\pi\,\sqrt{\{(\tfrac{1}{3}Ma^2+mx^2)/mgx\}}.$$

Hence the function of x whose variations are to be considered is

$$y=\frac{Ma^2}{3mx}+x.$$

The derivative is

$$\frac{dy}{dx}=-\frac{Ma^2}{3mx^2}+1.$$

Now since $a\geqslant x$ it follows that, if $M>3m$, dy/dx is negative; so that if the nut is raised slightly making dx negative, then dy is positive, or the period is increased. But if $M<3m$, then dy/dx is negative or positive according as $Ma^2>$ or $<3mx^2$, implying that the lengthening or shortening of the period for a slight displacement of the nut depends on its position.

15·54. Compound Pendulum with Axis Non-horizontal.
Suppose that a body is free to rotate about an axis that makes

an angle α with the vertical, the centre of gravity G will then oscillate in a plane making an angle α with the horizontal.

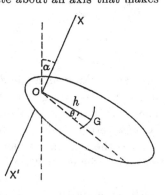

The weight mg can be resolved into $mg\sin\alpha$ parallel to the lines of greatest slope in this plane, and $mg\cos\alpha$ parallel to the axis. The latter component has no moment about the axis, and the former component alone is effective in causing oscillation.

Hence if GO is perpendicular to the axis and makes an angle θ with the line of greatest slope, and $GO=h$ and k is the radius of gyration about an axis through G parallel to the axis of rotation, the equation of moments about the fixed axis is

$$m\,(k^2+h^2)\,\ddot\theta=-mgh\sin\alpha\sin\theta,$$

so that the length of the equivalent simple pendulum is

$$(k^2+h^2)/h\sin\alpha,$$

and the period is

$$2\pi\,\sqrt{\{(k^2+h^2)/gh\sin\alpha\}}.$$

By making α small the period can be increased indefinitely.

15·55. Centre of Percussion. *If a single impulse can be applied to a rigid body which is free to turn about a fixed axis so as to produce no impulsive stress on the axis, the line of the impulse is called the line of percussion and the point in which this line meets the plane through the centre of gravity G and the axis of rotation is called the Centre of Percussion.*

We shall limit our investigation to the two-dimensional case, i.e. we suppose that the body is symmetrical about a plane through G perpendicular to the axis and that the impulse acts in this plane.

Let GO perpendicular to the axis be of length h, and let m be the mass and mk^2 the moment of inertia about a parallel axis through G. Suppose that when an impulse P is applied in a

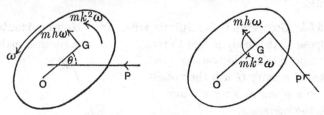

direction making an angle θ with GO there is no reaction at O and that the body begins to turn about O with angular velocity ω. The velocity of G is $h\omega$, so that the momentum is represented by $mh\omega$ perpendicular to GO and the spin couple $mk^2\omega$.

Since there is no reaction at O there can be no momentum set up in a direction at right angles to P, therefore

$$mh\omega \cos \theta = 0,$$

which requires that $\theta = \frac{1}{2}\pi$, or the direction of P must be at right angles to OG, as in the second figure. Let P meet OG at a distance p from O. Since there is no reaction at O, the equation of linear momentum is

$$mh\omega = P;$$

and, by moments about O, we get

$$m(h^2 + k^2)\,\omega = Pp.$$

Therefore $p = (h^2 + k^2)/h$, or the distance from O of the centre of percussion is equal to the length of the equivalent simple pendulum if the body were set to swing about the given axis placed horizontally.

15·6. Further examples of Motion about an Axis. (i) *The lock of a railway-carriage door will only engage if the angular velocity of the closing door exceeds ω. The door swings about vertical hinges and has a radius of gyration k about a vertical axis through the hinges, while the centre of gravity of the door is at a distance a from the line of the hinges. Shew that if the door be initially at rest and at right angles to the side of the train which then begins to move with acceleration f, the door will not close unassisted unless f > ½ ω²k²/a.* [M. T. 1920]

Let θ be the angle through which the door turns in time t. Let O represent the line of the hinges and let G be the centre of gravity. Then O has an acceleration f and G has accelerations $a\ddot\theta$ and $a\dot\theta^2$ relative to O, therefore the rates of change of momentum are, as in the figure, mf, $ma\dot\theta^2$, $ma\ddot\theta$ and $mk'^2\ddot\theta$, where k' is the

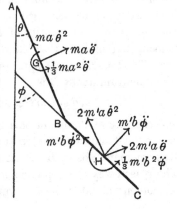

radius of gyration about a vertical line through G, so that $k^2 = a^2 + k'^2$. The only forces on the door in addition to its weight act through the line of the hinges, and therefore by taking moments about this line

$$m\,(a^2 + k'^2)\,\ddot\theta - maf\cos\theta = 0 \quad ....(1),$$

or

$$mk^2\ddot\theta - maf\cos\theta = 0.$$

Multiply by $2\dot\theta$ and integrate and we get

$$mk^2\dot\theta^2 - 2maf\sin\theta = C.$$

But $\dot\theta = 0$ when $\theta = 0$, since the door is initially at rest; therefore $C = 0$ and

$$k^2\dot\theta^2 = 2af\sin\theta.$$

When the door is about to close $\theta = \tfrac{1}{2}\pi$ and $\dot\theta^2 = 2af/k^2$. In order that the door may close we require that

$$\dot\theta > \omega, \quad \text{or} \quad f > \tfrac{1}{2}\omega^2 k^2/a.$$

(ii) *Motion of two heavy rods AB, BC smoothly jointed at B and swinging in a vertical plane about the end A which can turn about a fixed point.*

Let m, m' be the masses, and $2a$, $2b$ the lengths of the rods, and at time t let them make angles θ, ϕ with the vertical. Let G, H be their middle points. The accelerations of G are $a\dot\theta^2$ and $a\ddot\theta$ along and perpendicular to GA. The accelerations of H relative to B are $b\dot\phi^2$ and $b\ddot\phi$ along and perpendicular to HB, and B has accelerations $2a\dot\theta^2$ and $2a\ddot\theta$ along and perpendicular to BA which must be added to the two former components in order to get the total acceleration of H.

The rates of change of momentum for the rod AB are therefore $ma\dot\theta^2$, $ma\ddot\theta$, and the spin couple $\frac{1}{3}ma^2\ddot\theta$, and for the rod BC they are $m'b\dot\phi^2$, $m'b\ddot\phi$, $2m'a\dot\theta^2$, $2m'a\ddot\theta$, and the spin couple $\frac{1}{3}m'b^2\ddot\phi$. All these are shewn in the figure. The external forces, not shewn in the figure, consist of the reactions at A, B and the weights of the rods mg acting at G and $m'g$ acting at H.

In order to avoid introducing the unknown reactions at A and B we form equations by taking moments about B for the rod BC, and about A for the whole system.

By taking moments about B for the rod BC, we get

$$\frac{1}{3}m'b^2\ddot\phi + m'b^2\ddot\phi + 2m'ab\ddot\theta \cos(\phi-\theta)$$
$$+ 2m'ab\dot\theta^2 \sin(\phi-\theta) = -m'gb\sin\phi \quad\ldots\ldots\ldots(1).$$

And by taking moments about A for the two rods we get

$$\frac{1}{3}m'b^2\ddot\phi + m'b\ddot\phi\{b + 2a\cos(\phi-\theta)\} - m'b\dot\phi^2\, 2a\sin(\phi-\theta)$$
$$+ 2m'a\ddot\theta\{2a + b\cos(\phi-\theta)\} + 2m'a\dot\theta^2 b\sin(\phi-\theta) + \frac{1}{3}ma^2\ddot\theta + ma^2\ddot\theta$$
$$= -m'g(b\sin\phi + 2a\sin\theta) - mga\sin\theta \quad\ldots\ldots\ldots(2).$$

These two equations serve for the determination of the angular velocities; and one first integral might be found directly by writing down the equation of energy.

15·7. The last two examples serve to illustrate an important point. In establishing the theory of moment of momentum, see **9·2**, **14·1** and **14·3**, we always referred to moments about an axis fixed in space and the phrase employed was '*rate of change of moment of momentum.*' It is important however to note that, if the axis about which moments are taken is not fixed in space, then the phrases 'rate of change of moment of momentum' and 'moment of rate of change of momentum' are not equivalent. The latter phrase was used in **15·1**. Thus in the last example the components of velocity of H are $b\dot\phi$ relative to B and the velocity $2a\dot\theta$ of B, so that the components of momentum of BC are $m'b\dot\phi$, $2m'a\dot\theta$, and the spin couple $\frac{1}{3}m'b^2\dot\phi$. Hence the moment of momentum about B is

$$\tfrac{1}{3}m'b^2\dot\phi + m'b^2\dot\phi + 2m'a\dot\theta b\cos(\phi-\theta),$$

and the rate of change of this expression is not the same thing as the left-hand side of (1) in the last example, which represents 'the moment about B of the rate of change of momentum.' The reason why the expressions differ is because 'rate of change of moment about a *moving* point B' must take account of B's displacement during the short interval δt during which the change is observed, and is not the same as if the point B were fixed.

15·8. The student who desires to pursue this point further may revert to equation (4) of **9·2**, viz.

$$\Sigma m \, (x\ddot{y} - y\ddot{x}) = \Sigma \, (xY - yX).$$

This represents that the sum of the moments about the origin—*any* origin in the plane—of the rates of change of momentum ($m\ddot{x}$, $m\ddot{y}$) of the particles is equal to the sum of the moments about that origin of the external forces, and when in **15·6** (ii) we took the moments about B of the rates of change of momentum we were merely applying this principle and taking as origin the fixed point in the plane through which the junction B of the rods is passing at the instant considered.

But in **9·2** we go on to rewrite equation (4) in the form

$$\frac{d}{dt} \Sigma m \, (x\dot{y} - y\dot{x}) = \Sigma \, (xY - yX),$$

and read the left-hand side as 'rate of change of moment of momentum' instead of 'moment of rate of change of momentum,' and the important point is that we may not rewrite (4) in this way unless the point about which we take moments does not change its position while the operation d/dt is performed.

This will become clear if we suppose that at the instant considered the origin about which we take moments is moving with velocities u, v parallel to the axes. At time t the momenta of a particle m at (x, y) are $m\dot{x}$, $m\dot{y}$. At time $t + \delta t$ its coordinates have become $x + \dot{x}\,\delta t$, $y + \dot{y}\,\delta t$; but the origin has moved to $u\,\delta t$, $v\,\delta t$, so that the relative coordinates are

$$x + \dot{x}\,\delta t - u\,\delta t, \quad y + \dot{y}\,\delta t - v\,\delta t.$$

At the same time the momenta have become

$$m \, (\dot{x} + \ddot{x}\,\delta t), \quad m \, (\dot{y} + \ddot{y}\,\delta t),$$

so that the new moment of momentum about the moving origin is

$$\Sigma m \, \{(x + \dot{x}\,\delta t - u\,\delta t) \, (\dot{y} + \ddot{y}\,\delta t) - (y + \dot{y}\,\delta t - v\,\delta t) \, (\dot{x} + \ddot{x}\,\delta t)\},$$

and if we subtract the original moment of momentum

$$\Sigma m \, (x\dot{y} - y\dot{x})$$

and retain only the first power of δt, we get as the increment in the moment of momentum

$$\Sigma m \, (x\ddot{y} - y\ddot{x} - u\dot{y} + v\dot{x}) \, \delta t,$$

so that the rate of change of moment of momentum about the *moving* origin is

$$\Sigma m \, (x\ddot{y} - y\ddot{x}) - u\Sigma m\dot{y} + v\Sigma m\dot{x},$$

and this is not the same as

$$\frac{d}{dt} \Sigma m \, (x\dot{y} - y\dot{x}).$$

In conclusion therefore we repeat that it is only permissible to make use of 'rate of change of moment of momentum' when the axis about which moments are taken is fixed in space; but in every case it is legitimate to take moments about *any* axis and equate the moment of the rate of change of momentum and the moment of the external forces.

EXAMPLES

1. A reel of radius r with rims of radius R rests on a plane inclined at an angle α to the horizontal. A thread fixed to the reel passes round and under it and then upwards, parallel to the plane, and over a smooth pulley, a mass m hanging freely from this end. The thread lies in the vertical plane of symmetry of the reel, which is also perpendicular to the inclined plane. Calculate the tension on the thread and the acceleration of m (i) when the inclined plane is smooth, (ii) when the plane is rough and there is no slipping of the reel on the plane.   [M. T. 1917]

2. A reel of mass M, consisting of a cylinder of radius a connecting two discs of radius $b\,(b>a)$, rolls without slipping on a horizontal plane. A light thread, wound on the cylinder, passes in a plane perpendicular to the axis of the reel, horizontally from its under side, over a smooth pulley, and thence vertically downwards; to its free end is attached a mass m.

Shew that the reel will move with acceleration

$$f = g\,\frac{mb\,(b-a)}{M\,(b^2+k^2)+m\,(b-a)^2}\,,$$

where k is the radius of gyration of the wheel about its axis.   [M. T. 1927]

3. A uniform solid cylinder, mass M and radius a, rolls on a rough inclined plane with its axis perpendicular to the line of greatest slope. As it rolls the cylinder winds up a light string which passes over a fixed light pulley and supports a freely hanging mass m, the part of the string between the pulley and the cylinder being parallel to the lines of greatest slope. Discuss the motion of the cylinder, and prove that the tension of the string is

$$\frac{(3+4\sin\alpha)\,Mmg}{3M+8m}\,,$$

where α is the inclination of the plane to the horizontal.

[Coll. Exam. 1913]

4. A solid spherical ball rests in equilibrium at the bottom of a fixed spherical globe whose inner surface is perfectly rough. The ball is struck a horizontal blow of such magnitude that the initial speed of its centre is v. Prove that, if v lies between $\sqrt{10dg/7}$ and $\sqrt{27dg/7}$, the ball will leave the globe, d being the difference between the radii of the ball and globe.

[Coll. Exam. 1911]

5. A flywheel has a horizontal shaft of radius r; the moment of inertia of the system about the axis of revolution is K. A string of negligible thickness is wound round the shaft and supports a mass M hanging vertically. Find the angular acceleration of the wheel when its motion is opposed by a constant frictional couple G.

If the string is released from the shaft after the wheel has turned through an angle θ from rest, and if the wheel then turns through a further angle ϕ before it is brought to rest by the frictional couple, shew that

$$G = \frac{KMgr\theta}{K\theta + (K + Mr^2)\,\phi}.$$ [M. T. 1915]

6. A flywheel, turning with average angular velocity p, is acted on by a driving couple $A \sin^2 pt$, and has a constant couple $\frac{1}{2}A$ opposing its motion. Find the least moment of inertia required to make the difference between the greatest and least angular velocities less than $p/100$. [M. T. 1922]

7. A circular plate of mass M and radius a has a mass m fixed in it at a distance b from the centre. An axis through the centre of the plate and perpendicular to it can slide without friction horizontally, while the plate revolves. If the plate is just disturbed from rest when m is in its highest position, find the angular velocities when the disc has made one quarter and one half a turn.

Determine the pressure on the axis in each case. [M. T. 1918]

8. Shew that, if a uniform heavy right circular cylinder of radius a be rotated about its axis, and laid gently on two rough horizontal rails at the same level and distant $2a \sin \alpha$ apart so that the axis of the cylinder is parallel to the rails, the cylinder will remain in contact with both rails if the coefficient of friction $\mu < \tan \alpha$, but will initially rise on one rail if $\mu > \tan \alpha$. [M. T. 1919]

9. A uniform trap-door swinging about a horizontal hinge is closed by a spring coiled about the hinge. The spring is coiled so that it is just able to hold the door shut in the horizontal position. The horizontal opening which the door closes is in a body which is mounting with uniform acceleration f. Shew that, if $f = \left(0\cdot57 + \dfrac{1\cdot23}{\alpha}\right)g$, the door starting from the vertical position will just reach the horizontal position, α being the angle through which the spring is coiled when the door is in the horizontal position. [M. T. 1919]

10. An ellipse of axes a, b and a circle of radius b are cut from the same sheet of thin uniform metal and are superposed and fixed together with their centres coincident. The figure is free to move in its own vertical plane about one end of the major axis: shew that the length of the simple equivalent pendulum is $(5a^2 - ab + 2b^2)/4a$. [Coll. Exam. 1909]

11. A rigid pendulum OG swings about a horizontal axis through O, its centre of gravity being at G. The pendulum is released from rest when OG is horizontal. When OG is vertical, the pendulum is brought to rest by an inelastic buffer B which is such that the line of the reaction between B and the pendulum is horizontal and at a distance l below O. The mass of the pendulum is m, its moment of inertia about a horizontal axis through G is mk^2 and $OG = h$.

Shew that, if the impulse of the force exerted by B upon the pendulum during the impact is P,

$$lP = m\sqrt{2gh\,(h^2 + k^2)}.$$

Deduce the impulse Q of the horizontal force exerted on the pendulum during the impact by the axis O and shew that it vanishes when l is equal to the length of the equivalent simple pendulum.   [M. T. 1914]

12. A straight rod of mass m and length $2l$ swinging about one end as a compound pendulum starts from rest in a horizontal position and when vertical is struck a blow at its middle point which reverses and halves its angular velocity. Prove that the impulse of the blow $= m\sqrt{6gl}$.

[Coll. Exam. 1912]

13. A flywheel weighing 5 tons is suspended from a pair of centres entering conical holes in the rim so that it can swing in a vertical plane. The line joining the centres is parallel to and distant 4 feet from the axis of the wheel, and the period of a complete swing is 3·2 seconds. Find the radius of gyration of the wheel about its axis, and determine how much energy the wheel would lose in falling from a speed of 120 to 90 revolutions per minute when revolving round its axis.   [M. T. 1918]

14. A thin uniform rod of mass m and length $2a$ can turn freely about one end which is fixed, and a circular disc of mass $12m$ and radius $a/3$ can be clamped to the rod so that its centre is on the rod. Shew that, for oscillations in which the plane of the disc remains vertical, the length of the equivalent simple pendulum lies between $2a$ and $2a/3$.   [M. T. 1915]

15. A pendulum is supported at O, and P is the centre of oscillation. Shew that, if an additional weight is rigidly attached at P, the period of oscillation is unaltered.

16. A rope hangs over a pulley, which is of moment of inertia I, and perfectly smooth on its bearings, but perfectly rough to the rope. Two monkeys of equal mass m hang one on each end of the rope. The monkeys can climb with constant speeds u_1 and u_2 relative to the rope $(u_1 > u_2)$. Shew that in a race through a height h the monkey of speed u_1 can give the other monkey any start up to

$$hI\,(u_1 - u_2)/\{(I + ma^2)\,u_1 + ma^2 u_2\},$$

where a is the radius of the pulley. (The system is at rest before the monkeys start climbing.)   [S. 1925]

17. A fine string has masses M, M' $(M > M')$ attached to its ends and passes over a rough pulley with a fixed centre; shew that if the string does not slip, the downward acceleration of M is

$$g(M - M')/(M + M' + mk^2/a^2)$$

where m, mk^2, a are respectively the mass, moment of inertia about axis and radius of pulley.

Shew also that to prevent slip the coefficient of friction must be greater than

$$\frac{1}{\pi} \log \frac{2 + mk^2/(M'a^2)}{2 + mk^2/(Ma^2)}.$$ [M. T. 1918]

18. A uniform rod of mass m and length $2a$ is lying on a smooth horizontal table and is struck a blow B perpendicular to its length at one extremity. Find the velocities with which the two ends of the rod begin to move. [M. T. 1922]

19. Two uniform bars of the same material and cross-section, but of lengths a and b, are joined by a smooth hinge and laid out in a straight line. An impulse J at right angles to them is applied at the hinge. Shew that the hinge takes up a velocity $4J/m$, where m is the mass of the bars together. [Coll. Exam. 1927]

20. Two thin uniform rods AO, OB of lengths $2a$ and $2b$ and of masses m and M, are smoothly jointed at O, and lie in a straight line on a smooth table. An impulse P is applied at A perpendicular to AOB. Find the impulsive reaction at O and the initial velocity of O.

If equal and opposite impulses P are applied simultaneously at A and B, perpendicular to AOB, shew that the impulse at O is numerically equal to $\frac{1}{2}P$ and that the initial velocity of O is zero. [M. T. 1924]

21. Two flywheels, whose radii of gyration are in the ratio of their radii, are free to revolve in the same plane, a belt passing round both. Initially one, of mass m_1 and radius a_1, is rotating with angular velocity Ω, and the other, of mass m_2 and radius a_2, is at rest. Suddenly the belt is tightened, so that there is no more slipping at either wheel. Shew that the second wheel begins to revolve with angular velocity

$$\frac{m_1 a_1}{(m_1 + m_2) a_2} \Omega.$$ [S. 1923]

22. A circular disc of radius a lies on a smooth horizontal table, when a point on the circumference is compelled to move in the direction of the tangent at that point with velocity u. Prove that the disc begins to turn with angular velocity $2u/3a$. [Coll. Exam. 1910]

23. A rectangular lamina $ABCD$ of mass M and sides AB, CD of length $2a$, and AD, BC of length $2b$, rests upon a smooth horizontal plane. At a certain instant it receives a blow of impulse P applied at A and directed along AB. Shew that after a time t from this instant the distance of C from the original position of the centre of gravity is given by

$$\left\{ \frac{P^2 t^2}{M^2} + a^2 + b^2 - \frac{2P\sqrt{a^2+b^2}}{M}\, t \sin\left(\frac{3Pbt}{M(a^2+b^2)} - \epsilon \right) \right\}^{\frac{1}{2}},$$

where $\tan \epsilon = a/b$.   [M. T. 1926]

24. A uniform rod AB of mass M and length $2a$ is struck by an instantaneous impulse ξ, acting at B in direction BP which makes an angle θ with AB produced. Shew that AB will become parallel to BP after a time t given by

$$t = Ma\theta/3\xi \sin\theta.$$

If the end B, instead of being free, is compelled by a frictionless constraint to move in the line BP, shew that the impulse ξ will be accompanied by a simultaneous impulsive reaction on the constraint, of magnitude

$$3\xi \sin\theta \cos\theta/(1+3\cos^2\theta).$$ [M. T. 1927]

25. Two rods are smoothly jointed at A and B to a circular lamina, AB being a diameter of which the rods are continuations. The whole system lies at rest on a smooth horizontal table. Shew that a horizontal blow P applied normally to the rim of the lamina will produce in the latter no initial angular velocity, provided that

$$\frac{m_1 k_1^2}{a_1^2 + k_1^2} = \frac{m_2 k_2^2}{a_2^2 + k_2^2},$$

where m_1, m_2 are the masses of the rods, a_1, a_2 are the distances of their centres of gravity from A and B respectively, and k_1, k_2 are their moments of inertia about their centres of gravity.   [Coll. Exam. 1911]

26. A free lamina of any form is turning in its own plane about an instantaneous centre of rotation S. A point P in the line joining S to the centre of gravity G is brought to rest by an impulsive force passing through the point. Find the position of P in terms of SG and the radius of gyration about G, assuming that the velocity of G is the same as before but reversed in direction.   [M. T. 1918]

27. If a body can only turn about a smooth horizontal axle, and when the body is at rest, the axle is given an instantaneous horizontal velocity v in a direction perpendicular to its length, prove that the centre of mass will start off with a velocity $v(k^2 - h^2)/k^2$, and that the initial angular velocity will be vh/k^2, where h is the distance of the centre of mass from the axle and k the radius of gyration about the axle.   [Coll. Exam. 1913]

28. A gate of length l can swing about a vertical hinge at one end. When it is in equilibrium a particle of mass m and velocity v strikes the gate normally at its middle point. The moment of inertia of the gate about the hinge is $I = \frac{1}{8}ml^2$, and the coefficient of restitution is $\frac{1}{2}$. Find the velocity of the particle and the angular velocity of the gate just after impact. [M. T. 1928]

29. The speed of a railway truck, weighing 5 tons, is reduced uniformly from 25 to 20 miles per hour on the level in a distance of $695\frac{3}{4}$ ft. by the brakes. Shew that, if no slipping takes place between the wheels and the rails, the normal pressure between the rails and each of the front wheels is 50 lb. weight greater than the corresponding pressures on the back wheels, given that the distance between the axles is 12 ft. and that the centre of gravity of the truck is $4\frac{1}{2}$ ft. above the ground and equidistant from the axles, while the diameter of each wheel is 3 ft. and the moment of inertia of each pair of wheels and axle about its axis is 3600 lb. ft.^{2} units. [M. T. 1916]

30. A thin uniform rod OA, of length a and mass m, turns in a horizontal plane about a vertical axis through O; an equal rod AB is jointed at A to OA and a smooth guide compels the end B to move along a horizontal straight line Ox. The angle $AOx = \theta$. No external forces act upon the rods except those due to the axis and the guide, and the motion takes place without friction. When $\theta = 0$, $d\theta/dt = \omega$. Shew that

$$(d\theta/dt)^2 = \omega^2/(4 - 3\cos^2\theta).$$

Prove that, if H be the angular momentum of the system about the vertical axis through O,

$$H = ma^2\, d\theta/dt.$$

Shew that the force which the smooth guide Ox exerts upon the rod AB is

$$3ma\omega^2 \sin\theta/2\,(4 - 3\cos^2\theta)^2. \qquad \text{[M. T. 1923]}$$

31. A rectangular gate is free to swing about one edge which is inclined at an angle of 5° to the vertical. When the gate is shut the top and bottom lines are horizontal. The gate is then opened through 90° and slightly disturbed so that it shuts. Find its angular velocity in any position, and the time occupied in the last 45° of its swing. The gate is uniform and 4 feet wide. [M. T. 1924]

ANSWERS

1. (i) $T = g\,(1 + \sin\alpha)\Big/\Big(\dfrac{1}{m} + \dfrac{1}{M} + \dfrac{r^2}{Mk^2}\Big);\ f = g - \dfrac{T}{m}.$

(ii) $T = g\,\{k^2 + R^2 + R\,(R - r)\sin\alpha\}\Big/\Big\{\dfrac{k^2 + R^2}{m} + \dfrac{(R - r)^2}{M}\Big\};\ f = g - \dfrac{T}{m}.$

5. $(Mgr - G)/(K + Mr^2)$.   6. $50A/p^2$.

7. $2\{mgb/(2mb^2 + Ma^2)\}^{\frac{1}{2}}$; $2\{2m(M+m)gb/[M(M+m)a^2 + 2Mmb^2]\}$.

$(M+m)g - 2m^2b^2g/Ma^2$;

$$(M+m)g\{1 + 8m^2b^2g/[M(M+m)a^2 + 2Mmb^2]\}.$$

11. $Q = \dfrac{mh}{l}\left(l - \dfrac{k^2 + h^2}{h}\right)\sqrt{\left(\dfrac{2gh}{k^2 + h^2}\right)}$.   13. $4 \cdot 09$ ft. ; 205408 ft. lb.

18. $4B/m$, $-2B/m$.   20. $MP/2(M+m)$; $2P/(M+m)$.

26. $k^2(GP - SG) = 2SG \cdot GP^2$.   28. $\frac{1}{2}v$; $2v/l$.

31. $8\{2\sin 5^\circ(1 - \cos\theta)\}^{\frac{1}{2}}/k$; $\dfrac{k}{8\sqrt{\sin 5^\circ}}\log_e\dfrac{\tan 22^\circ 30'}{\tan 11^\circ 15'}$ seconds where θ is the angle turned through and k is the radius of gyration about the hinges.

Chapter XVI

MISCELLANEOUS PROBLEMS

16·1. Rolling and Sliding. When sliding takes place the friction F bears a constant ratio μ to the normal reaction R but when rolling takes place the friction has generally a much smaller value.

When two bodies in contact at a point A have a relative motion the process of determining whether this motion involves rolling or sliding at A is as follows:

Write down the equations of motion which involve the friction F and the normal reaction R, and assuming (i) that rolling takes place also write down the kinematical condition which expresses the fact that there is no relative tangential velocity at A. If from the solution of these equations we find that F/R is less than the coefficient of friction μ, the assumption that rolling takes place is justified and rolling will continue until F/R becomes greater than μ.

(ii) Assuming that sliding takes place, write μR instead of F in the equations of motion and solve the equations without the kinematical condition above referred to. If the solution shews that there is a relative tangential velocity at A and the direction of motion is opposed to what has been assumed as the direction of the friction then we have found the true motion and it will continue until relative velocity at A vanishes.

16·12. *A wheel spinning about a horizontal axis is projected along a rough horizontal plane, to determine the subsequent motion.*

Let m be the mass and a the radius of the wheel, mk^2 its moment of inertia about its centre G, and A the point of contact with the plane.

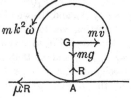

Suppose that initially G has velocity v_0 and that there is an angular velocity ω_0 in the opposite sense to what it would be if the motion were one of pure rolling. Let v, ω be what these velocities become at time t. Since the point A of the wheel has a velocity $v + a\omega$, sliding is taking place and the friction is μR where R is the vertical reaction.

The equations of motion are

$$m\dot{v} = -\mu R, \quad 0 = R - mg, \quad mk^2\dot{\omega} = -\mu Ra.$$

Therefore

$$\dot{v} = -\mu g \quad \text{and} \quad k^2\dot{\omega} = -\mu ga.$$

Hence, by integrating and putting in the initial values,

$$v = v_0 - \mu gt \quad \text{and} \quad k^2\omega = k^2\omega_0 - \mu gat \quad\ldots\ldots\ldots\ldots\ldots(1).$$

These equations shew that v and ω decrease steadily and the subsequent motion depends on which vanishes first.

v vanishes after a time $v_0/\mu g$ and ω vanishes after a time $k^2\omega_0/\mu ga$. Let us suppose (i) that v vanishes first, i.e. that $v_0 < k^2\omega_0/a$. At the instant when v vanishes the angular velocity has the value ω_1 given, from (1), by

$$k^2\omega_1 = k^2\omega_0 - av_0$$

or

$$\omega_1 = \omega_0 - av_0/k^2 \quad\ldots\ldots\ldots\ldots\ldots\ldots\ldots\ldots(2),$$

which is positive in the same sense as ω_0, so that the point A still has a velocity $a\omega_1$ in the same sense, i.e. there is still slipping at A though G has come to rest and there is still friction μR in the same sense.

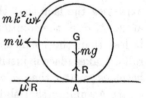

This friction will now give G an acceleration $\dot{u}$ in the reversed direction, and the equations of motion while sliding lasts are

$$m\dot{u} = \mu R, \quad 0 = R - mg, \quad mk^2\dot{\omega} = -\mu Ra$$

or

$$\dot{u} = \mu g \quad \text{and} \quad k^2\dot{\omega} = -\mu ga$$

so that $u = \mu gt$ and $k^2\omega = k^2\omega_1 - \mu gat \;\ldots(3)$, where t is the time since G was at rest.

These equations shew that u increases and ω decreases, and the velocity of the point A is

$$u - a\omega = \mu gt\left(1 + \frac{a^2}{k^2}\right) - a\omega_1 \quad\ldots\ldots\ldots\ldots\ldots\ldots(4).$$

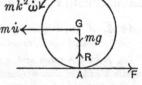

This velocity vanishes after a time $k^2a\omega_1/\mu g\,(k^2 + a^2)$ when sliding ceases, and since at that time u is $k^2a\omega_1/(k^2 + a^2)$ therefore the wheel rolls, and the equations become

$$m\dot{u} = -F, \quad mk^2\dot{\omega} = Fa$$

and the condition for rolling $u - a\omega = 0$.

These equations shew that $\dot{u} = 0$ or $u = \text{constant} = k^2a\omega_1/(k^2 + a^2)$ where ω_1 is given by (2), and the wheel continues to roll with this constant velocity.

Reverting now to the first stage of the motion and suppose alternatively (ii) that $av_0 > k^2\omega_0$ so that ω vanishes first. This happens, from (1), when $t = k^2\omega_0/\mu ga$, and at that instant the point of contact A has the same velocity as G, namely, $v_1 = v_0 - k^2\omega_0/a$ from (1), so that sliding continues and the friction μR begins to create an angular velocity Ω in the opposite sense to the former angular velocity. Measuring t from the instant at which ω vanishes, the equations of motion are

$$m\dot{v} = -\mu R, \quad 0 = R - mg, \quad mk^2\dot{\Omega} = \mu Ra$$

or $\qquad \dot{v} = -\mu g, \quad k^2\dot{\Omega} = \mu ga.$

Therefore, by integrating and using the initial values, we get

$$v = v_1 - \mu gt \quad \text{and} \quad k^2\Omega = \mu gat.$$

The velocity of A is

$$v - a\Omega = v_1 - \mu gt\,(1 + a^2/k^2)$$

which vanishes when

$$t = v_1/\mu g\,(1 + a^2/k^2).$$

Sliding then ceases, and since v is not then zero rolling begins and as before it may be shewn that the final velocity is uniform.

16·2. Two Spheres in Contact. *A sphere of mass m and radius a rolls on a sphere of mass m' and radius b which rolls on a horizontal plane; the friction is sufficient to prevent sliding and the motion is all parallel to the same vertical plane.*

Let A, B be the centres of the spheres and ω, ω' their angular velocities. Let v denote the velocity of the sphere m', and let BA make an angle θ with the vertical at time t.

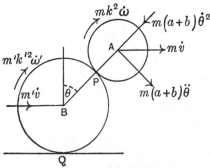

Relative to B, A is describing a circle of radius $a+b$, therefore its accelerations relative to B are $(a+b)\,\ddot{\theta}$ at right angles to AB and $(a+b)\,\dot{\theta}^2$ along AB, and with these must be compounded the acceleration $\dot{v}$ of B. Hence the rates of change of momentum of the sphere m are $m\dot{v}$, $m\,(a+b)\,\ddot{\theta}$, $m\,(a+b)\,\dot{\theta}^2$ and the spin couple $mk^2\dot{\omega}$ or $\frac{2}{5}ma^2\dot{\omega}$. Similarly the rates of change of momentum of m' are $m'\dot{v}$ and the spin couple $m'k'^2\dot{\omega}'$ or $\frac{2}{5}m'b^2\dot{\omega}'$.

There are two kinematical conditions for rolling, viz. the point of contact Q of the lower sphere with the plane has no velocity, therefore

$$v - b\omega' = 0 \quad \dotfill (1).$$

Also the point of contact P of the two spheres must have no relative tangential velocity, or the tangential velocities of the point P on the two spheres must be the same.

On the sphere m, P has velocity $a\omega$ relative to A, but A's velocity is $(a+b)\,\dot{\theta}$ relative to B and B's velocity is v; therefore on the sphere m, P's tangential velocity is

$$(a+b)\,\dot{\theta}+v\cos\theta-a\omega.$$

But on the sphere m', P has velocity $b\omega'$ relative to B so that its tangential velocity is $b\omega'+v\cos\theta$; therefore

$$(a+b)\,\dot{\theta}-a\omega=b\omega' \quad\dots\dots\dots\dots\dots\dots\dots\dots\dots(2).$$

We have now to write down the equations of motion and we may avoid the unknown reactions by taking moments about P for the sphere m, and about Q for the two spheres together. Thus we have

$$ma\,(a+b)\,\ddot{\theta}+ma\dot{v}\cos\theta+\tfrac{2}{5}ma^2\dot{\omega}=mga\sin\theta \quad\dots\dots\dots(3),$$

and $\quad m\,(a+b)\,\ddot{\theta}\,(a+b+b\cos\theta)-m\,(a+b)\,\dot{\theta}^2 b\sin\theta+m\dot{v}\,\{(a+b)\cos\theta+b\}$

$$+\tfrac{2}{5}ma^2\dot{\omega}+m'b\dot{v}+\tfrac{2}{5}m'b^2\dot{\omega}'=mg\,(a+b)\sin\theta\dots\dots\dots(4).$$

And equations (1), (2), (3) and (4) are sufficient to determine the motion.

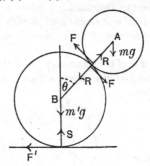

We might alternatively have written down the equation of energy for the whole system. Observing that the velocity of A is compounded of v and $(a+b)\,\dot{\theta}$, this equation is

$$\tfrac{1}{2}m\,\{v^2+(a+b)^2\,\dot{\theta}^2+2\,(a+b)\,\dot{\theta}v\cos\theta+\tfrac{2}{5}a^2\omega^2\}$$

$$+\tfrac{1}{2}m'\,(v^2+\tfrac{2}{5}b^2\omega'^2)+mg\,(a+b)\cos\theta=\text{const}\dots\dots(5).$$

This equation is of course a first integral of the equations of motion.

If however we wish to determine the unknown reactions, denoting them as in the second figure, we can write down the formal equations of motion for each sphere; resolving, and taking moments about the centre for each sphere we have

$$m\,(a+b)\,\ddot{\theta}+m\dot{v}\cos\theta=mg\sin\theta-F,$$

$$m\,(a+b)\,\dot{\theta}^2-m\dot{v}\sin\theta=mg\cos\theta-R,$$

$$\tfrac{2}{5}ma^2\dot{\omega}=Fa,$$

$$m'\dot{v}=-F'-R\sin\theta+F\cos\theta,$$

$$0=S-m'g-R\cos\theta-F\sin\theta,$$

$$\tfrac{2}{5}m'b^2\dot{\omega}'=F'a+Fa.$$

These six equations together with (1) and (2) are equivalent to (1), (2), (3) and (4) and also determine the unknown reactions.

16·21. *The same problem as the last but no friction between the lower sphere and the plane.*

In this case equation (1) no longer holds, but instead we have the condition that there is no external horizontal force acting on the system, so that if it starts from rest its horizontal momentum remains zero throughout the motion. This fact is expressed by the equation

$$mv + m(a+b)\,\dot\theta\cos\theta + m'v = 0,$$

and the other equations are as in the last article.

16·3. Initial Motions and Stresses.

When one or more of the constraints which maintain a system in equilibrium are removed the initial motion and the initial values of the remaining stresses in the system may be determined by writing down the kinematical conditions that specify the remaining constraints together with the equations of motion of the system in its initial position, with the simplification that initial velocities are all zero so that radial and transverse accelerations are simply $\ddot r$ and $r\ddot\theta$ and all normal components of acceleration vanish.

Examples. (i) *A heavy uniform straight rod is suspended from a point by two strings of the same length as the rod attached to its ends. If one string is cut, prove that the initial angular acceleration of the rod is nine times that of the remaining string.*

Let m be the mass and $2a$ the length of the rod; and let $\dot\omega$, $\dot\omega'$ denote the initial accelerations of the string and rod.

The middle point G of the rod has acceleration $a\dot\omega'$ relative to the end A; but the end A of the string has acceleration $2a\dot\omega$ perpendicular to the string OA, therefore G has also this acceleration, and the rates of change of momentum are $ma\dot\omega'$, $2ma\dot\omega$ and the spin couple $\tfrac{1}{3}ma^2\dot\omega'$.

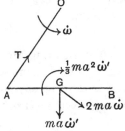

If T be the initial tension in the string, by resolving horizontally for the rod, we get

$$2ma\dot\omega\cos 30^\circ = T\cos 60^\circ \quad\quad\quad\quad\quad\quad(1),$$

and by taking moments about G

$$\tfrac{1}{3}ma^2\dot\omega' = Ta\sin 60^\circ \quad\quad\quad\quad\quad\quad(2),$$

whence we get

$$\dot\omega' = 9\omega \quad\quad\quad\quad\quad\quad\quad\quad\quad(3).$$

To find T in terms of the weight of the rod we resolve vertically and get

$$ma\dot\omega' + 2ma\dot\omega\cos 60^\circ = mg - T\sin 60^\circ,$$

and with the help of (3) and (1) we find that $T = \dfrac{2\sqrt{3}}{13}\,mg.$

(ii) *A solid hemisphere is held with its base against a smooth vertical wall and its lowest point on a smooth floor. The hemisphere is released. Find the initial pressures on the wall and the floor.*

Let m be the mass, a the radius, C the centre and G the centre of gravity of the hemisphere. $CG = \frac{3}{8}a$. Let O be the point of contact with the ground and let $OG = h$.

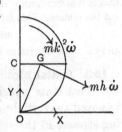

Since the hemisphere begins to turn about O, if $\dot{\omega}$ is the initial angular acceleration, the acceleration of G is $h\dot{\omega}$ at right angles to OG. Hence the rate of change of momentum is represented by $mh\dot{\omega}$ and the spin couple $mk^2\dot{\omega}$ where k is the radius of gyration about a horizontal axis through G parallel to the wall.

By taking moments about the line of intersection of the wall and floor we get

$$m(h^2 + k^2)\,\dot{\omega} = \tfrac{3}{8}mga.$$

Now $m(h^2 + k^2)$ is the moment of inertia of the hemisphere about the line of intersection of the wall and floor. But its moment of inertia about a diameter of its face is $\frac{2}{5}ma^2$, and therefore by the theorem of parallel axes (**13·2**) the moment of inertia about the required axis is $\frac{7}{5}ma^2$.

Hence $\frac{7}{5}ma^2\dot{\omega} = \frac{3}{8}mga$, or $a\dot{\omega} = \frac{15}{56}g$.

Let X, Y be the pressures of the wall and floor on the hemisphere, then by resolving horizontally and vertically we get

$$mh\dot{\omega}\cos COG = X, \quad \text{or} \quad ma\dot{\omega} = X,$$
and $$mh\dot{\omega}\sin COG = mg - Y \quad \text{or} \quad \tfrac{3}{8}ma\dot{\omega} = mg - Y.$$

Therefore

$$X = \tfrac{15}{56}mg \quad \text{and} \quad Y = \tfrac{403}{448}mg.$$

16·4. Bending Moments in Bodies in Motion. When a rod AB is in equilibrium under the action of given coplanar forces the stresses at any point P of the rod may be determined by imagining the rod to be divided by a cross-section at P. Then

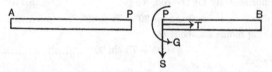

by considering the equilibrium of either portion, say PB, we see that the forces exerted by AP on PB must balance all the other given forces that act on PB. But a system of coplanar forces can be reduced to a single force acting at any assigned point in the plane together with a couple. Hence the given forces acting on PB can be represented by a force at P with a tangential component T and a normal component S together with a couple G, and the reaction of AP on PB balances these. T is called the tension, S the shearing force and G the bending moment, and the latter is taken as the measure of the tendency of the rod to break at P.

When the rod is in a given state of motion instead of in equilibrium, if we write down the equations of motion for either portion these equations will suffice for the determination of T, S and G, if the accelerations have first been determined by considering the motion of the whole rod.

16·41. Example. *A heavy rod swings freely in a vertical plane about one end; to find the stresses at any point of the rod.*

Let l be the length of the rod OA, m the mass of unit length, and θ its inclination to the vertical at time t.

By taking moments about the fixed end O for the whole rod, we get

$$\tfrac{1}{3}ml^3\ddot\theta = -\tfrac{1}{2}mgl^2\sin\theta$$

or $\qquad l\ddot\theta = -\tfrac{3}{2}g\sin\theta \ldots\ldots\ldots\ldots(1),$

and, by multiplying by $2\dot\theta$ and integrating,

$$l\dot\theta^2 = \tfrac{3}{2}g(\cos\theta - \cos\alpha) \ldots\ldots(2),$$

when α denotes the amplitude of the oscillation.

Now let the point P at which the stresses are required be at a distance a from O, and let T, S, G denote the tension, shearing force and bending moment in the sense in which they act on PA. We have to write down the equations of motion of the part PA of the rod. Consider a small element of length dx and mass mdx at Q where $OQ = x$. The accelerations of this element are $x\ddot\theta$ and $x\dot\theta^2$ at right angles to and along QO. Hence by resolving at right angles to and along AO and taking moments about P, for the portion PA of the rod, we get

$$\int_a^l mx\ddot\theta\,dx = S - mg(l-a)\sin\theta \ldots\ldots\ldots\ldots\ldots\ldots(3),$$

$$\int_a^l mx\dot\theta^2\,dx = T - mg(l-a)\cos\theta \ldots\ldots\ldots\ldots\ldots\ldots(4),$$

and $\qquad \int_a^l mx\ddot\theta\,(x-a)\,dx = -G - \tfrac{1}{2}mg(l-a)^2\sin\theta \ldots\ldots\ldots(5).$

These are equivalent to

$$\tfrac{1}{2}m\,(l^2-a^2)\,\ddot{\theta}=S-mg\,(l-a)\sin\theta\ \dotfill(3)',$$

$$\tfrac{1}{2}m\,(l^2-a^2)\,\dot{\theta}^2=T-mg\,(l-a)\cos\theta\ \dotfill(4)',$$

and $$\tfrac{1}{6}m\,(l-a)^2\,(2l+a)\,\ddot{\theta}=-G-\tfrac{1}{2}mg\,(l-a)^2\sin\theta\dotfill(5)'.$$

Substituting for $\ddot{\theta}$ from (1) and for $\dot{\theta}^2$ from (2) we get

$$S=\tfrac{1}{4}mg\,\frac{(l-a)\,(l-3a)}{l}\sin\theta,$$

$$T=mg\,(l-a)\cos\theta+\tfrac{3}{4}mg\,\frac{(l^2-a^2)}{l}(\cos\theta-\cos a),$$

and $$G=\tfrac{1}{4}mg\,\frac{(l-a)^2}{l}\,a\sin\theta.$$

16·5. Steady Motion in Three Dimensions.

A steady motion is one in which all velocity components are constant in magnitude. Problems of steady motion are often solved by simple applications of the principles of energy and momentum.

Examples. (i) *A uniform straight rod of length l is free to turn about one end. Find the angular velocity with which it can describe a cone of semi-vertical angle a.*

If ω is the angular velocity then each element of the rod describes a horizontal circle with angular velocity ω.
Let m be the mass of unit length and consider an element mdx at a distance x from the fixed end O. Its acceleration is $\omega^2 x\sin a$, therefore its rate of change of momentum is $mdx\,.\,\omega^2 x\sin a$ towards the centre of the circle that it describes.

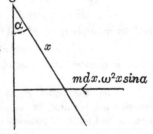

Take moments for the whole rod about a horizontal axis through O perpendicular to the rod, and we get

$$\int_0^l x\cos a\,.\,m\omega^2 x\sin a\,dx=mlg\,.\,\tfrac{1}{2}l\sin a,$$

or $$\tfrac{1}{3}m\omega^2\,l^3\cos a\sin a=\tfrac{1}{2}mgl^2\sin a,$$

therefore $$l\omega^2\cos a=\tfrac{3}{2}g.$$

(ii) *The framework in the figure revolves about the vertical axis and the balls move outwards for an increase of speed, the weight E sliding up the axis. Shew that the angular velocity ω with which the frame rotates in the position in which the arm ABD makes an angle a with the vertical is given by*

$$(a+l\sin a)\,\omega^2=(1+m'b/ml)\,g\tan a,$$

where $$AB=BC=b,\quad AD=l,\quad AA'=CC'=2a.$$

The balls are each of mass m and are fixed to the arms: the sliding weight E is of mass m' and the weights of the arms may be neglected.   [M. T. 1917]

The point D describes a horizontal circle of radius $a + l \sin a$ with angular velocity ω. Therefore the acceleration of D is $(a + l \sin a) \omega^2$ towards the axis of rotation and the rate of change of momentum of the ball is $m (a + l \sin a) \omega^2$. We have now to consider whether to resolve or take moments and we notice that we cannot resolve without introducing

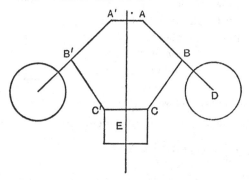

the stresses in the rod AD to which the ball is fixed, and we cannot take moments about any point of the rod except the point A without introducing the bending moment of the rod. We therefore take moments for the ball and rod AD about an axis through A perpendicular to the plane of the diagram. This gives

$$m (a + l \sin a) \omega^2 . l \cos a = mgl \sin a + Tb \sin 2a,$$

where T is the tension in the rod BC.

Next resolving vertically for the weight E, which is at rest in the steady motion, we have

$$2T \cos a = m'g.$$

Eliminating T, we get

$$(a + l \sin a) \omega^2 = (1 + m'b/ml) g \tan a.$$

16·6. Use of the Instantaneous Centre of Rotation. In the motion of a body in one plane, let u, v denote the velocities of its centre of gravity G parallel to rectangular axes and let ω be the angular velocity of the body. If (r, θ) are the polar coordinates of a point P of the body referred to G as origin, the velocity of P relative to G is $r\omega$ at right angles to GP and this has components $-r\omega \sin \theta$, $r\omega \cos \theta$ parallel to 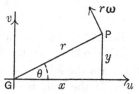 the axes of x and y or $-y\omega$, $x\omega$ respectively, if (x, y) are coordinates of P relative to the axes through G. Therefore the whole velocities of P are

$$u - y\omega \quad \text{and} \quad v + x\omega$$

and if these components vanish the point P has no velocity. Therefore the point whose coordinates are $-v/\omega$, u/ω is at rest at the instant considered. This point I is called the *instantaneous centre of rotation* (5·41).

Let us now write down the equation of motion for the body obtained by taking moments about I. The rates of change of momentum are $m\dot{u}$, $m\dot{v}$ and the spin couple $mk^2\dot{\omega}$, and if L denotes the sum of the moments about I of the external forces we have

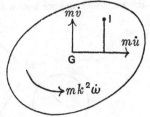

L = moment about I of rate of change of momentum

$$= m\dot{u}\,\frac{u}{\omega} - m\dot{v}\left(-\frac{v}{\omega}\right) + mk^2\dot{\omega}$$

$$= \frac{m}{\omega}\,(u\dot{u} + v\dot{v} + k^2\omega\dot{\omega})$$

$$= \frac{1}{\omega}\,\frac{d}{dt}\,\{\tfrac{1}{2}m\,(u^2 + v^2 + k^2\omega^2)\}.$$

If now r denotes the distance GI, since the body is, at the instant considered, turning about I, the velocity of G is $r\omega$, so that $u^2 + v^2 = r^2\omega^2$, and therefore

$$\frac{1}{2\omega}\,\frac{d}{dt}\,\{m\,(r^2 + k^2)\,\omega^2\} = L \quad\ldots\ldots\ldots\ldots(1).$$

By the theorem of parallel axes (13·2) if K denotes the radius of gyration about I, we have

$$K^2 = r^2 + k^2,$$

and therefore

$$\frac{1}{2\omega}\,\frac{d}{dt}\,(mK^2\omega^2) = L \quad\ldots\ldots\ldots\ldots\ldots(2).$$

Since the instantaneous centre is generally changing its position relative to the body, K is not generally constant but there are the following special cases:

(i) If the body is turning about a fixed axis K is constant and (2) becomes

$$mK^2\dot{\omega} = L \quad\ldots\ldots\ldots\ldots\ldots\ldots\ldots(3),$$

as in 15·5, and the same is true if the instantaneous centre is at a constant distance from G.

(ii) *If the axis be not fixed but the body starts from rest*, since (2) is equivalent to

$$mK^2\dot{\omega} + \tfrac{1}{2}\omega m\,\frac{dK^2}{dt} = L\ldots\ldots\ldots\ldots\ldots(4),$$

and initially $\omega = 0$, therefore the *initial* value of the acceleration is given by

$$mK^2\dot{\omega} = L\ \ldots\ldots\ldots\ldots\ldots\ldots\ldots(3).$$

(iii) In *a small oscillation* about a position of equilibrium, if we take moments about the instantaneous centre in a position slightly displaced from the equilibrium position, then in equation (4) ω is small and dK^2/dt is of the order of a velocity, and therefore the second term of the equation is of the order of the square of a small velocity and can be neglected, so that again

$$mK^2\dot{\omega} = L\ \ldots\ldots\ldots\ldots\ldots\ldots\ldots(3).$$

To summarize the results, it appears that equation (3) is a valid equation if the instantaneous axis of rotation is a fixed axis or at a fixed distance from the centre of gravity, or if we are dealing with an initial motion or a small oscillation; but in every other case in which we take moments about the instantaneous centre or axis of rotation we must use equation (2).

16·61. Examples. (i) *The ends of a heavy rod are constrained to move on two smooth intersecting wires, one of which is vertical and the other horizontal.*

Let AB be the rod, $2a$ its length, m its mass and θ its inclination to the vertical. The instantaneous centre of rotation is the corner I of the rectangle $OAIB$ (**5·42**) where Ox, Oy are the wires.

Also, since $GI = a$, the moment of inertia of the rod about I

$$= m\left(\tfrac{1}{3}a^2 + GI^2\right)\ \ \textbf{(13·2)}$$
$$= \tfrac{4}{3}ma^2.$$

This is a case therefore in which we can employ equation (3) of **16·6**, taking moments about I and thereby avoiding the unknown reactions at A and B which pass through I. The angular velocity of the rod is $\dot{\theta}$ in the counter-clockwise sense, and the only force that has a moment about I is the weight mg in the same sense, therefore

$$\tfrac{4}{3}ma^2\ddot{\theta} = mga\sin\theta.$$

On multiplying by $\dot{\theta}$ and integrating, this gives

$$\tfrac{2}{3}a\dot{\theta}^2 = g\,(\cos\alpha - \cos\theta),$$

where α is the initial inclination of the rod to the vertical.

We leave it to the student to obtain the same result by the principle of energy.

(ii) *The figure represents the cross-section through the centre of a uniform elliptic cylinder which rests on a smooth horizontal plane against a smooth vertical plane. The minor axis of the cross-section passes through the intersection O of the planes. To find the initial angular acceleration.*

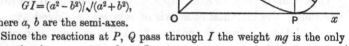

The normals at the points of contact P, Q meet in the instantaneous centre I, and in the symmetrical position OGI is a straight line, where G is the centre of gravity, and it can be shewn that

$$GI = (a^2 - b^2)/\sqrt{(a^2 + b^2)},$$

where a, b are the semi-axes.

Since the reactions at P, Q pass through I the weight mg is the only force that has a moment about I.

Therefore if $\dot{\omega}$ be the initial angular acceleration

$$m\,(k^2 + GI^2)\,\dot{\omega} = mgGI/\sqrt{2}.$$

But

$$k^2 = (a^2 + b^2)/4,$$

therefore

$$(5a^4 - 6a^2b^2 + 5b^4)\,\dot{\omega} = 2\sqrt{2}g\,(a^2 - b^2)\,\sqrt{(a^2 + b^2)}.$$

16·62. In the next chapter we shall have occasion to use the instantaneous centre of rotation in examples of small oscillations.

EXAMPLES

1. A circular hoop of radius a, while spinning in a vertical plane with angular velocity ω about its centre, is gently placed on a rough inclined plane which slopes at the angle of friction α for the surfaces in contact. Shew that, if the sense of the rotation be that which causes the slipping at the point of contact to be down the line of greatest slope, the hoop will remain stationary for a time $a\omega/g\sin\alpha$.

2. A uniform sphere, of radius a, is projected with velocity V down a rough slope of inclination α having also an angular velocity Ω about a horizontal axis in such a sense as to tend to cause rolling up a line of

greatest slope. Prove that it will never stop slipping unless $\mu > \frac{2}{7} \tan \alpha$ and that, if $5V < 2a\Omega$, it will turn back if $\mu > \tan \alpha \Big/ \left(1 - \dfrac{5}{2} \dfrac{V}{a\Omega}\right)$.

[Coll. Exam. 1928]

3. A circular hoop in a vertical plane is projected down an inclined plane with velocity V_0 and at the same time is given an angular velocity ω_0 tending to make it roll up the plane. Find the relation between V_0, ω_0, the slope of the plane and the coefficient of friction, if the hoop comes to a position of instantaneous rest.   [Coll. Exam. 1914]

4. A circular cylinder is fixed with its axis horizontal, and a sphere is placed on the highest generator and slightly disturbed. If the surfaces are rough, find the position of the sphere when sliding begins, and shew that the sphere must slide before leaving the cylinder.   [Coll. Exam. 1910]

5. A homogeneous sphere of radius r is placed on a smooth horizontal table and a perfectly rough equal homogeneous sphere is placed on the top and then slightly displaced. Prove that the same points always remain in contact and that the angular velocity of either sphere is $\sqrt{(5g/6r)}$ at the instant of impact with the table.   [Coll. Exam. 1909]

6. Investigate the motion of a circular hoop, hanging over a rough peg whose cross-section is circular, and find the components of the reaction in any position, assuming that the peg is sufficiently rough to prevent slipping.   [Coll. Exam. 1908]

7. A sphere of mass m rolls down the rough face of an inclined plane of mass M and angle α, which is free to slide on a smooth horizontal plane in a direction perpendicular to its edge. Investigate the motion and shew that the pressure between the sphere and the inclined plane is

$$m\,(2m + 7M)\, g \cos \alpha / \{(2 + 5 \sin^2 \alpha)\, m + 7M\}.$$

[Coll. Exam. 1912]

8. A sphere of radius a and mass M rests on a smooth horizontal plane, and a second sphere of radius b and mass m is placed upon it, the line of centres being inclined at an angle θ_0 with the vertical; the surfaces between the two spheres are rough. The system is allowed to move from rest in this position. Shew that as long as the spheres remain in contact

$$(7M + 5m \sin^2 \theta)\,(a+b)\,\dot{\theta}^2 = 10\,(M+m)\,g\,(\cos \theta_0 - \cos \theta),$$

where θ is the inclination of the line of centres to the vertical.

[Coll. Exam. 1927]

9. A perfectly rough solid sphere, of mass m and radius r, rests symmetrically upon a hollow cylinder, of mass M and radius R, free to turn about its axis which is horizontal. If the sphere roll down, shew that at any time during the contact the angle ϕ between the line of centres and the vertical is given by

$$(7M+2m)(R+r)\ddot{\phi}=(5M+2m)g\sin\phi\,;$$

and find the value of ϕ when the bodies separate.   [Coll. Exam. 1926]

10. The mass of a sphere is $\frac{1}{5}$ of that of another sphere of the same material, which is free to move about its centre as a fixed point, and the first sphere rolls down the second from rest at the highest point, the coefficient of friction being μ. Prove that sliding will begin when the angle θ which the line of centres makes with the vertical is given by

$$\sin\theta=2\mu(5\cos\theta-3).\qquad\text{[M. T. 1905]}$$

11. A rough perfectly elastic sphere is dropped without rotation on to a horizontal cylinder which is free to turn about its axis. There is no slipping at the point of contact during the impact, and the sphere starts moving horizontally after the impact. If θ is the angle which the radius of the cylinder through the point of contact makes with the vertical, prove that

$$\tan^2\theta=1+\cfrac{1}{\dfrac{ma^2}{I}+\dfrac{ma'^2}{I'}},$$

where a and a' are the radii of the cylinder and sphere, I and I' are their moments of inertia about their centres and m is the mass of the sphere.

Shew also that the coefficient of friction between the cylinder and sphere must be greater than $\frac{1}{2}(\tan\theta-\cot\theta)$ in order that there may be no slipping during the impact.   [M. T. 1922]

12. A uniform straight heavy rod AB of mass M is freely jointed about a smooth horizontal axis at A, and is supported at an inclination θ to the vertical by a light string which is perpendicular to the rod and attached to it at B. The string is suddenly cut. Find the pressures on the axis at A before and immediately afterwards.   [M. T. 1919]

13. A square lamina is suspended by vertical strings tied to two adjacent corners; two edges of the square being vertical. If one string is cut, shew that the tension of the remaining string is instantaneously diminished to $\frac{4}{5}$ of its former value.   [M. T. 1911]

14. A uniform rod is held at inclination α to the vertical, with its lower end resting on an imperfectly rough table. It is suddenly released. Shew that the lower end will instantly begin to slide if the coefficient of friction is less than $\dfrac{3\sin\alpha\cos\alpha}{4-3\sin^2\alpha}$, and hence that if the coefficient of friction exceeds $\frac{3}{4}$ the rod will not slide initially whatever α may be.

[Coll. Exam. 1927]

15. A uniform heavy rectangular plate is placed with an edge along the intersection of a smooth vertical wall and a smooth horizontal plane and allowed to turn round the edge from a position of rest in a nearly vertical position: shew that the edge will leave the wall during the motion.

[Coll. Exam. 1915]

16. A uniform circular disc of radius a is rolling without slipping along a smooth horizontal plane with velocity V when the highest point becomes suddenly fixed. Prove that the disc will make a complete revolution round the point if $V^2 > 24ag$. [M. T. 1928]

17. A heavy uniform rod of mass m is supported against a smooth fixed sphere by a horizontal string fastened to its upper end and also to the highest point of the sphere. Shew that, if the string is cut, the initial pressure of the rod on the sphere is $mg\cos\alpha/(1+3\sin^4\alpha)$, where α is the angular distance of the point of contact from the highest point of the sphere.

18. A smooth ring of mass m slides on a wire bent into the form of a circle of radius r which is made to rotate about a vertical diameter with uniform angular velocity ω. Find the position of relative rest of the ring on the wire and shew that the pressure between the ring and the wire is then $m\omega^2 r$. [S. 1917]

19. Two unequal masses are connected by a string of length l which passes through a fixed smooth ring. The smaller mass moves as a conical pendulum while the other mass hangs vertically. Find the semi-angle of the cone, and the number of revolutions per second when a length a of the string is hanging vertically. [S. 1927]

20. The ends of a uniform rod can slide without friction on a circular wire which is made to rotate about a vertical diameter with angular velocity ω. Shew that, if there is a state of steady motion in which the rod is not horizontal, its inclination to the horizontal is $\cos^{-1}(g/a\omega^2)$, where a is the distance of its centre from the centre of the circle.

21. A rigid ring hanging over a smooth peg is set spinning about its centre (which remains stationary) in its own vertical plane. It is completely fractured at one point A. Prove that the maximum bending moment experienced at the diametrically opposite point A' is

$$2r^3\omega^2 + gr^2(4+\pi^2)^{\frac{1}{2}},$$

and find the corresponding direction of AA'. The ring has unit mass per unit length, its radius is r and its angular velocity is ω. [M. T. 1918]

22. The ends of a uniform rod of length l can slide freely along two smooth wires at right angles to one another, one of which is horizontal and the other vertical and below the first. Shew that, if the frame rotates round the vertical wire with uniform angular velocity ω, there is a position of relative equilibrium for the rod, in which its lower end is at a depth $3g/2\omega^2$ below the horizontal wire.

23. A rectangular lamina of diagonal $2b$ can move in a vertical plane with adjacent sides in contact with two smooth pegs at a distance a apart in the same horizontal line. Prove that the angular velocity $\dot{\theta}$ is given by

$$\{a^2 + \tfrac{4}{3}b^2 - 2ab\cos(\alpha - \theta)\}\,\dot{\theta}^2 + g\{2b\sin(\alpha + \theta) - a\sin 2\theta\} = C,$$

where α is the inclination of the diagonal to a side and θ is the inclination of that side to the horizontal.

ANSWERS

3. $\mu(a\omega_0 - V_0) = a\omega_0 \tan\alpha$, where a is the radius and α the slope.
4. $2\sin\theta = \mu(17\cos\theta - 10)$.   6. The equivalent simple pendulum is of length equal to the difference of the diameters of the hoop and peg. The reactions when the line of centres makes an angle θ with the vertical are $mg(2\cos\theta - \cos\alpha)$, $\tfrac{1}{2}mg\sin\theta$; where m is the mass and α the extreme value of θ.   12. The reaction along the rod remains $Mg\cos\theta$, the reaction at right angles to the rod changes from $\tfrac{1}{2}Mg\sin\theta$ to $\tfrac{1}{4}Mg\sin\theta$.
18. The radius to the ring makes an angle $\cos^{-1}(\omega^2 r/g)$ with the vertical.
19. $\cos^{-1}(m/m')$; $\{m'g/4\pi^2 m(l-a)\}^{\frac{1}{2}}$, where m, m' are the masses.
21. AA' makes an angle $\tan^{-1}\tfrac{1}{2}\pi$ with the vertical.

Chapter XVII

SMALL OSCILLATIONS

17·1. In Chapter VII we considered the case of harmonic oscillations of finite amplitude of a particle moving in a straight line, the characteristic equation for which is

$$\ddot{x} + n^2 x = 0,$$

and we remarked (**7·23**) that an equation of this form always represents a periodic motion with a period independent of the amplitude. There is a large class of problems wherein the equation of motion can be reduced to this form, though generally with the limitation that x remains small throughout the motion, this class includes all problems of a system having one degree of freedom which is slightly disturbed from a position of stable equilibrium and proceeds to oscillate about that position. The usual method of solving such a problem is to write down the equation of energy, i.e. express the fact that the sum of the kinetic and potential energies is constant, then differentiate this equation. The method will be explained more fully in the next article.

17·2. Application of the Principle of Energy. If a system possesses only one degree of freedom its position can be defined by the values of a single variable x, which may denote a linear or an angular displacement. Let us suppose that the system has a position of stable equilibrium and that the origin from which x is measured is so chosen that x vanishes in the position of equilibrium. In any other position the kinetic energy T must be proportional to $\dot{x}^2$, say

$$T = \tfrac{1}{2} A \dot{x}^2 \quad \dots\dots\dots\dots\dots\dots(1),$$

where A may contain x but does not vanish when x vanishes; and the potential energy V will be a function of x, which we assume can be expanded in ascending powers of x in the form

$$V = V_0 + ax + bx^2 + cx^3 + \dots \quad \dots\dots\dots\dots(2).$$

Now in a position of stable equilibrium the potential energy V is a minimum*, therefore $dV/dx = 0$ when $x = 0$, and this requires that $a = 0$. Further the oscillations that we are about to consider are assumed to be so small that powers of x and $\dot{x}$ above the second power can be neglected. This means that in (1), if A is a function of x, since A is multiplied by $\dot{x}^2$, it will be sufficient for our purpose to retain the part of A that is independent of x, or put $x = 0$ in A; and in (2), since $a = 0$, we may now write

$$V = V_0 + bx^2 \quad\dots\dots\dots\dots\dots\dots(3),$$

neglecting higher powers of x than the second.

The equation of energy $T + V = $ const. now becomes

$$\tfrac{1}{2}A\dot{x}^2 + bx^2 = \text{const.} \quad\dots\dots\dots\dots(4),$$

where A and b are constants.

Differentiate with regard to t and divide by $\dot{x}$ and we get

$$A\ddot{x} + 2bx = 0 \quad\dots\dots\dots\dots\dots\dots(5),$$

which represents a periodic motion of period $2\pi \sqrt{(A/2b)}$.

17·3. Examples. (i) *A solid uniform circular cylinder of mass m and of radius r rolls (under the action of gravity) inside a fixed hollow cylinder of radius R, the axes of the cylinders being parallel to each other and also horizontal. At any time t during the motion the plane containing the axes of the cylinders makes an angle θ with the vertical. Shew that the potential energy of the moving cylinder is mg (R − r) (1 − cos θ), and that its kinetic energy is $\tfrac{3}{4} m (R-r)^2 \dot{\theta}^2$. Hence, or otherwise, shew that the time, T, of a small oscillation is*

$$T = 2\pi \sqrt{\{3\,(R-r)/2g\}}. \qquad \text{[M. T. 1914]}$$

Let C and G be the axes of the fixed and rolling cylinders, and let O be the equilibrium position of the point G. Then $CO = CG = R - r$; and the potential energy is the weight multiplied by its height above O, i.e. $(CO - CG \cos\theta)$, therefore

$$V = mg\,(R - r)\,(1 - \cos\theta),$$

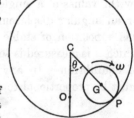

and, as far as the second power of θ, this is

$$V = \tfrac{1}{2}mg\,(R - r)\,\theta^2.$$

Again, let ω denote the angular velocity of the rolling cylinder, then the point of contact P has a velocity $r\omega$ relative to the axis G, but G has a velocity $(R-r)\dot{\theta}$ in the opposite direction, so that the condition that P has no velocity at the instant considered gives

$$r\omega = (R-r)\dot{\theta} \quad\dots\dots\dots\dots\dots\dots(1).$$

* See *Note on the Energy Test of Stability*, p. 257.

Since the velocity of the centre of gravity G is $(R-r)\,\dot{\theta}$, and the moment of inertia of the cylinder about its axis is $\tfrac{1}{2}mr^2$, therefore the kinetic energy is

$$\tfrac{1}{2}m\{(R-r)^2\,\dot{\theta}^2+\tfrac{1}{2}r^2\omega^2\}=\tfrac{3}{4}m\,(R-r)^2\,\dot{\theta}^2.$$

The equation of energy is therefore

$$\tfrac{3}{4}m\,(R-r)^2\,\dot{\theta}^2+\tfrac{1}{2}mg\,(R-r)\,\theta^2=\text{const.},$$

when higher powers of θ are neglected.

Differentiating and dividing by $m\,(R-r)\,\dot{\theta}$, we get

$$\tfrac{3}{2}(R-r)\,\ddot{\theta}+g\theta=0 \dotfill (2).$$

This represents oscillatory motion with period

$$2\pi\,\sqrt{\{3\,(R-r)/2g\}}.$$

Otherwise, the accelerations of G are $(R-r)\,\ddot{\theta}$ and $(R-r)\,\dot{\theta}^2$ at right angles to and along GC; therefore the rates of change of momentum are $m\,(R-r)\,\ddot{\theta}$, $m\,(R-r)\,\dot{\theta}^2$, and the spin couple $\tfrac{1}{2}mr^2\dot{\omega}$. Therefore by taking moments about the line of contact of the cylinders we get

$$m\,(R-r)\,r\ddot{\theta}+\tfrac{1}{2}mr^2\dot{\omega}=-mgr\sin\theta,$$

but from (1) $r\dot{\omega}=(R-r)\,\ddot{\theta}$, therefore the equation reduces to

$$\tfrac{3}{2}(R-r)\,\ddot{\theta}=-g\sin\theta,$$

which, if we write θ for $\sin\theta$, thus neglecting θ^3, is the same as equation (2).

(ii) *The ends of a uniform rod of length $2a$ can slide on a smooth circular wire of radius b in a vertical plane. If the rod makes small oscillations about its equilibrium position, find the length of the equivalent simple pendulum.*

Let AB be the rod, m its mass and G its centre; and let O be the centre of the wire. At time t let OG make a small angle θ with the vertical. Then $OG=\sqrt{(b^2-a^2)}$, and the potential energy is

$mg\,OG\,(1-\cos\theta)$, or $\tfrac{1}{2}mg\,\sqrt{(b^2-a^2)}\,\theta^2$,

to the second power of θ.

Again, the velocity of G is $OG\dot{\theta}$ or

$$\sqrt{(b^2-a^2)}\,\dot{\theta},$$

and the angular velocity of the rod is $\dot{\theta}$, therefore the kinetic energy is

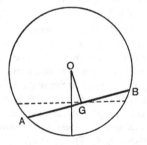

$$\tfrac{1}{2}m\{(b^2-a^2)\,\dot{\theta}^2+\tfrac{1}{3}a^2\dot{\theta}^2\}=\tfrac{1}{6}m\,(3b^2-2a^2)\,\dot{\theta}^2.$$

The equation of energy is therefore

$$\tfrac{1}{6}m\,(3b^2-2a^2)\,\dot{\theta}^2+\tfrac{1}{2}mg\,\sqrt{(b^2-a^2)}\,\theta^2=\text{const.},$$

when higher powers of θ are neglected.

Differentiating and simplifying, we get

$$\tfrac{1}{3}(3b^2-2a^2)\,\ddot{\theta}+g\,\sqrt{(b^2-a^2)}\,\theta=0.$$

Therefore the length of the equivalent simple pendulum is

$$(3b^2-2a^2)/3\,\sqrt{(b^2-a^2)}.$$

17·4. Use of the Instantaneous Centre of Rotation. We saw in **16·6** that in problems on small oscillations we can take moments about the instantaneous centre of rotation as though it were a fixed point, writing

$$mK^2\dot{\omega} = L,$$

where mK^2 is the moment of inertia and L the moment of the external forces about the axis of rotation through the instantaneous centre.

As an application consider the last example of **17·3.**

In this case O is the instantaneous centre and the moment of inertia of the rod AB about O is by the theorem of parallel axes (**13·2**)

$$m\left(\frac{a^2}{3}+GO^2\right)=m\left(\frac{a^2}{3}+b^2-a^2\right)=\tfrac{1}{3}m\,(3b^2-2a^2).$$

Also the reactions at A and B pass through O and therefore have no moment about O, and the moment of the weight is

$$mg\,OG\sin\theta,\quad\text{or}\quad mg\,\sqrt{(b^2-a^2)}\sin\theta.$$

Hence, neglecting θ^3, we have

$$\tfrac{1}{3}m\,(3b^2-2a^2)\,\ddot{\theta}=-mg\,\sqrt{(b^2-a^2)}\,\theta,$$

as in **17·3.**

17·41. Example. *Two uniform rods each of length 2a and of the same mass m are smoothly jointed and placed over two smooth pegs in the same horizontal line, so that in equilibrium the rods make equal angles α with the vertical. Find the period of the small oscillations in which the joint moves vertically.*

Let P, Q be the pegs at a distance $2c$ apart, and let G be the centre of gravity of a rod AB, A being the joint. Since when A moves vertically the rod AB slides over the peg P, therefore the instantaneous centre I for the rod AB is where the perpendicular to the rod through P meets the horizontal through A. The reaction at P and the reaction of the rod AC on AB are along PI and IA respectively. Therefore in equilibrium the vertical through G goes through I, and

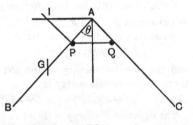

$$a\sin\alpha=IA=AP\cosec\alpha=c\cosec^2\alpha,$$

or

$$c=a\sin^3\alpha \quad\dots\dots\dots\dots\dots\dots\dots(1).$$

In a slightly displaced position let θ be the inclination of the rods to the vertical. The vertical through G no longer passes through I, and by taking moments about I for the rod AB we have

$$m\left(\tfrac{1}{3}a^2 + GI^2\right)\ddot{\theta} = -mg\left(a\sin\theta - c\,\mathrm{cosec}^2\,\theta\right).$$

But $GI^2 = GP^2 + PI^2 = (a - c\,\mathrm{cosec}\,\theta)^2 + c^2\cot^2\theta\,\mathrm{cosec}^2\,\theta$

$$= a^2 - 2ac\,\mathrm{cosec}\,\theta + c^2\,\mathrm{cosec}^4\,\theta.$$

Therefore

$$\left(\tfrac{1}{3}a^2 - 2ac\,\mathrm{cosec}\,\theta + c^2\,\mathrm{cosec}^4\,\theta\right)\ddot{\theta} = -g\left(a\sin\theta - c\,\mathrm{cosec}^2\,\theta\right) \quad ...(2).$$

Here θ is not small, but differs from the constant α by a small quantity, so that we may write $\theta = \alpha + \chi$ where χ is small and the first power only of χ is to be retained in the equation. Since now $\ddot{\theta} = \ddot{\chi}$, therefore, in the coefficient of $\ddot{\chi}$ on the left-hand side of the equation, it will be a sufficiently close approximation to take $\theta = \alpha$. The right-hand side is a function of θ which by (1) vanishes when θ is put equal to α. To obtain the value of this expression correct to the first power of χ, the simplest method is to remember that, if χ^2 can be neglected,

$$f(\theta) = f(\alpha + \chi) = f(\alpha) + \chi f'(\alpha),$$

but by (1) $f(\alpha) = 0$, therefore $f(\theta) = \chi f'(\alpha)$.

In this case $f(\theta) = -g\left(a\sin\theta - c\,\mathrm{cosec}^2\,\theta\right),$

therefore $f'(\theta) = -g\left(a\cos\theta + 2c\,\mathrm{cosec}^3\,\theta\cos\theta\right),$

and $f'(\alpha) = -g\left(a\cos\alpha + 2c\,\mathrm{cosec}^3\,\alpha\cos\alpha\right)$

$$= -3ga\cos\alpha, \text{ from (1)}.$$

Hence the right-hand side of (2) becomes $-3ga\cos\alpha\,.\,\chi$ and we have

$$\left(\tfrac{1}{3}a^2 - 2ac\,\mathrm{cosec}\,\alpha + c^2\,\mathrm{cosec}^4\,\alpha\right)\ddot{\chi} = -3ga\cos\alpha\,.\,\chi,$$

or, by substituting from (1),

$$\left(\tfrac{1}{3} - \sin^2\alpha\right)a\ddot{\chi} + 3g\cos\alpha\,.\,\chi = 0,$$

or $\left(1 + 3\cos^2\alpha\right)a\ddot{\chi} + 9g\cos\alpha\,.\,\chi = 0,$

making the period

$$2\pi\,\sqrt{\{a\left(1 + 3\cos^2\alpha\right)/9g\cos\alpha\}}.$$

17·5. Oscillations of a Particle constrained to move on a Revolving Curve. Let a particle P of mass m be constrained to move on a plane curve which revolves with uniform angular velocity ω about an axis in its plane; the field of force being the same in all planes through this axis. Take the axis of rotation for axis of y and a perpen-

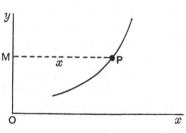

dicular for axis of x. Let the coordinates of the particle be x, y

at time t. Let X, Y be the components parallel to the axes of the forces acting on the particle, including the reaction of the curve. The effect of the rotation of the curve is to make the x component of acceleration of the particle $\ddot{x} - x\omega^2$ instead of $\ddot{x}$; because, if PM be perpendicular to the axis of y, the radius vector MP is turning about M in a plane perpendicular to Oy with angular velocity ω, so that the radial component of acceleration (12·1) is $\ddot{x} - x\omega^2$. Hence the equations of motion are

$$m(\ddot{x} - x\omega^2) = X, \quad m\ddot{y} = Y,$$

or $$m\ddot{x} = X + m\omega^2 x, \quad m\ddot{y} = Y.$$

Therefore the effect of the uniform rotation can be provided for in the solution of the problem by adding to the forces acting on the particle a force $\omega^2 x$ per unit mass directed from the axis of rotation and otherwise regarding the given curve as at rest.

17·51. Example. *A particle is constrained to move on a smooth circular wire which rotates uniformly about a diameter which is vertical. If in the position of relative rest the radius drawn to the particle makes an angle α with the vertical, find the period of small oscillations about this position.*

Let a be the radius of the circle, m the mass of the particle and let θ be the angle that the radius drawn to the particle in any position makes with the vertical. We may regard the wire as at rest if we add to the forces acting on the particle a force $m\omega^2 a \sin\theta$ away from the vertical diameter. The other forces are the weight mg and the normal reaction. Hence in relative rest

$$\tan\theta = m\omega^2 a \sin\theta / mg,$$

or, since $\theta = \alpha$ in relative rest, therefore $a\omega^2 = g \sec\alpha$. And in any general position by resolving along the tangent

$$ma\ddot{\theta} = m\omega^2 a \sin\theta \cos\theta - mg \sin\theta.$$

For small oscillations about the position $\theta = \alpha$, put $\theta = \alpha + \chi$ where χ is small, and we get

$$a\ddot{\chi} = \tfrac{1}{2}\omega^2 a \sin(2\alpha + 2\chi) - g \sin(\alpha + \chi)$$
$$= \tfrac{1}{2}\omega^2 a (\sin 2\alpha + 2\chi \cos 2\alpha) - g(\sin\alpha + \chi \cos\alpha).$$

Then, substituting $g \sec\alpha$ for $a\omega^2$, we have

$$a\ddot{\chi} + g\chi \sin^2\alpha / \cos\alpha = 0,$$

which gives a period

$$2\pi \sqrt{(a \cos\alpha / g \sin^2\alpha)}.$$

17·6. Stability of Steady Motion. In previous chapters and in the last two articles we have had examples of steady motion, some such were described in **16·5**. Such a motion is said to be stable if when owing to any cause the motion is slightly disturbed it continues without a wide departure from the original path.

Suppose, for example, that a particle is describing a circle of radius $1/u_0$ under the action of a force to the centre which would produce an acceleration $f(u)$ at distance $1/u$.

Let h be the moment of the velocity about the centre in the circular orbit, then hu_0 is the velocity, and the acceleration $f(u_0)$ is (vel.)^2/(radius), i.e. $h^2 u_0{}^3$, or

$$h^2 = f(u_0)/u_0{}^3.$$

Now suppose that without altering h, so that $1/u_0$ remains an apsidal distance, the particle is slightly displaced from the circular orbit. The equation for the path with the law of force $f(u)$ is, by **12·21**,

$$\frac{d^2 u}{d\theta^2} + u = \frac{f(u)}{h^2 u^2},$$

or

$$\frac{d^2 u}{d\theta^2} + u = \frac{u_0{}^3 f(u)}{u^2 f(u_0)}.$$

Putting $u = u_0 + x$, where x is small, and retaining only the first power of x, we get

$$\frac{d^2 x}{d\theta^2} + u_0 + x = \frac{u_0{}^3 f(u_0 + x)}{f(u_0)(u_0 + x)^2}$$

$$= u_0 \left\{ 1 + \frac{x f'(u_0)}{f(u_0)} \right\} \left(1 - \frac{2x}{u_0} \right)$$

Therefore

$$\frac{d^2 x}{d\theta^2} + x \left\{ 3 - \frac{u_0 f'(u_0)}{f(u_0)} \right\} = 0.$$

If the coefficient of x is positive and we denote it by n^2, we have

$$\frac{d^2 x}{d\theta^2} + n^2 x = 0,$$

the solution of which is $x = A \cos(n\theta + \alpha)$.

This shews that x is periodic in θ and oscillates in its values so that the motion is stable. Also, if x vanishes for a given value of θ, it will next vanish when θ is increased by π/n, so that the apsidal angle or angle between consecutive apses is π/n, i.e.

$$\pi \Big/ \left\{ 3 - \frac{u_0 f'\,(u_0)}{f\,(u_0)} \right\}^{\frac{1}{2}}$$

The condition for stability is that the expression under the square root must be positive. If the force varies as the kth power of the distance, i.e. $f(u) \propto \dfrac{1}{u^k}$, $u_0 f'\,(u_0)/f(u_0) = -k$, so that $3 + k$ must be positive and k must not be less than -3.

When the force varies as the distance, $k = 1$, and the apsidal angle $= \frac{1}{2}\pi$. The disturbed circular orbit becomes in fact an ellipse with the centre of force in the centre and there are four apses at the ends of the axes (**12·41**).

When the force varies inversely as the square of the distance, $k = -2$, and the apsidal angle $= \pi$. In this case the disturbed orbit is an ellipse with the centre of force in a focus and there are two apses at the ends of the major axis (**12·5**).

17·7. The usual method of procedure in solving a problem on disturbed steady motion is first to write down the equations of motion in their general forms for the given system under the given field of force together with the kinematical conditions, if any. Next deduce from these equations the condition for steady motion by making all the second differential coefficients zero. This means that we are making all velocities constant and finding the particular value of the variable, $\theta = \alpha$ say, which defines the position in steady motion. Next suppose the motion to be slightly disturbed from the steady state and substitute in the original equations $\theta = \alpha + \chi$ where χ is small, then by retaining only the first power of χ and its derivatives we obtain the equation between $\ddot{\chi}$ and χ, which gives the period of the oscillations.

We shall illustrate this process by some examples.

17·8. *The point of suspension of a pendulum is A, and A is caused to move along a horizontal straight line OX. The centre of gravity of the pendulum is G, and AG = l. The radius of gyration about any axis through G perpendicular to AG is k. The pendulum can move in the vertical plane* *containing OX. At time t, OA=x, and the angle between AG and the vertical is θ, supposed positive when GAX is obtuse. Prove that*

$$(l^2 + k^2)\,\ddot{\theta} + lg \sin\theta = l\cos\theta \,.\, \ddot{x}.$$

If ẍ has the constant value f, shew that the pendulum can maintain a constant inclination a to the vertical, where tan a = f/g, and that the periodic time of small oscillations about this position is

$$2\pi \left\{ \frac{l^2 + k^2}{lg} \cos a \right\}^{\frac{1}{2}}. \qquad \text{[M. T. 1924]}$$

The point A has acceleration $\ddot{x}$, and G has accelerations $l\ddot{\theta}$ and $l\dot{\theta}^2$ relative to A at right angles to and along GA. Therefore, if m be the mass of the pendulum, the rates of change of momentum are $m\ddot{x}$, $ml\ddot{\theta}$, $ml\dot{\theta}^2$, and the spin couple $mk^2\ddot{\theta}$.

Hence by taking moments about A we get

$$ml^2\ddot{\theta} - m\ddot{x}l\cos\theta + mk^2\ddot{\theta} = -mgl\sin\theta,$$

or $(l^2 + k^2)\,\ddot{\theta} + lg\sin\theta = l\cos\theta \,.\, \ddot{x}$ ....(1).

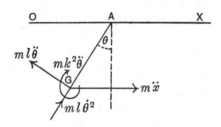

This is the general equation of motion, and if $\ddot{x}$ has the constant value f, it becomes

$$(l^2 + k^2)\,\ddot{\theta} + lg\sin\theta = lf\cos\theta.......................(2).$$

In this case we find the possible steady motion by putting $\ddot{\theta} = 0$, which gives $\tan\theta = f/g$; and we denote this value of θ by a. Substituting $g\tan a$ for f in (2) gives

$$(l^2 + k^2)\,\ddot{\theta} + lg\sin(\theta - a)/\cos a = 0 \;.................(3).$$

To find the period of the small oscillations about this position we now put $\theta = a + \chi$, where χ is small, and neglect all powers of χ above the first. Hence we get

$$(l^2 + k^2)\,\ddot{\chi} + lg\chi/\cos a = 0,$$

giving a period $2\pi \left\{ \dfrac{l^2 + k^2}{lg} \cos a \right\}^{\frac{1}{2}}.$

It should be observed that the latter part of this question can be solved by elementary considerations, for the relative effect on the pendulum of giving its point of suspension a constant horizontal acceleration is equivalent to keeping the point of suspension fixed and applying a force mf to the pendulum at G in the opposite sense to the acceleration. The equilibrium position of AG is then the direction of the resultant of f and g, and its time of oscillation is got from the ordinary formula for the period of a compound pendulum by writing $\sqrt{(g^2+f^2)}$ or $g\sec\alpha$ instead of g.

17·9. *A particle moving on the inside of a smooth fixed sphere describes a horizontal circle. Shew that if ω be the angular velocity the depth of the circle below the centre of the sphere is g/ω^2, and find the period of small oscillations about this steady motion.*

Let m be the mass of the particle and a the radius of the sphere. Let the radius to the particle make an angle θ with the downward vertical, and let the meridian circle make an angle ϕ with a fixed vertical plane.

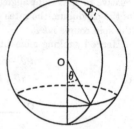

In the general motion of the particle on the sphere its velocity along the vertical circle is $a\dot\theta$ and along the horizontal circle $a\sin\theta\,.\,\dot\phi$. Therefore the equation of energy is

$$\tfrac{1}{2}ma^2\,(\dot\theta^2+\sin^2\theta\dot\phi^2)+mga\,(1-\cos\theta)$$
$$=\text{const.}\dots(1).$$

Again the particle is acted on by its weight mg, and the reaction of the surface which passes through O, neither of which has any moment about the vertical diameter. Therefore the moment of momentum about the vertical diameter is constant, i.e.

$$ma^2\sin^2\theta\,.\,\dot\phi=\text{const.}$$

Hence if there be a steady motion in which $\theta=\alpha$ and $\dot\phi=\omega$,

$$\sin^2\theta\,.\,\dot\phi=\omega\sin^2\alpha\dots\dots\dots\dots\dots\dots\dots(2).$$

Eliminating $\dot\phi$ between (1) and (2), we get

$$\tfrac{1}{2}a\dot\theta^2+\tfrac{1}{2}a\omega^2\sin^4\alpha\,\mathrm{cosec}^2\,\theta-g\cos\theta=\text{const.}$$

By differentiating this equation it follows that

$$a\ddot\theta-a\omega^2\sin^4\alpha\,\mathrm{cosec}^3\,\theta\cos\theta+g\sin\theta=0\dots\dots\dots(3).$$

The condition for steady motion with $\theta=\alpha$ is got by putting $\ddot\theta=0$, whence

$$a\omega^2\cos\alpha=g,$$

or the depth of the horizontal circle below the centre of the sphere is

$$a\cos\alpha=g/\omega^2\dots\dots\dots\dots\dots\dots\dots\dots(4).$$

Substitute $g\sec\alpha$ for $a\omega^2$ in (3) and we get

$$a\ddot\theta-g\sin^4\alpha\sec\alpha\,\mathrm{cosec}^3\,\theta\cos\theta+g\sin\theta=0\dots\dots\dots(5).$$

Then for small oscillations about the horizontal circle we put $\theta=\alpha+\chi$, where χ is small, and retain the first power of χ only.

Now $f(\theta)=f(\alpha+\chi)=f(\alpha)+\chi f'(\alpha),$

and to get $f'(\alpha)$ we first differentiate $f(\theta)$ and then put $\theta=\alpha$.

In this case $f(\theta) = -g \sin^4 \alpha \sec \alpha \operatorname{cosec}^3 \theta \cos \theta + g \sin \theta$,

and $f'(\theta) = 3g \sin^4 \alpha \sec \alpha \operatorname{cosec}^4 \theta \cos^2 \theta + g \sin^4 \alpha \sec \alpha \operatorname{cosec}^2 \theta + g \cos \theta$.

Therefore $f(\alpha) = 0$, and

$$f'(\alpha) = 4g \cos \alpha + g \sin^2 \alpha \sec \alpha$$
$$= g (1 + 3 \cos^2 \alpha) \sec \alpha.$$

Hence (5) becomes

$$a\ddot{\chi} + g (1 + 3 \cos^2 \alpha) \sec \alpha . \chi = 0,$$

and the period of the small oscillations is

$$2\pi \sqrt{\{a \cos \alpha / g (1 + 3 \cos^2 \alpha)\}}.$$

Note on the Energy Test of Stability. The proposition that positions of minimum and maximum potential energy are positions of stable and unstable equilibrium respectively is generally found in books on Statics; it is however an application of the principle of conservation of energy.

In the first place if the potential energy of a system is stationary in value for small permissible displacements from an assigned position, then no work is done by the external forces in any such displacement so that, by the principle of virtual work, such a position is one of equilibrium.

Now consider a body in a position of minimum potential energy. If the body undergoes a small but finite displacement in any direction compatible with its geometrical connections and is then set free so that motion ensues, it begins to acquire kinetic energy and therefore to lose potential energy. It moves therefore so that its potential energy decreases, i.e. back towards the position of minimum potential energy, and that position is therefore one of stable equilibrium. In like manner it can be shewn that a position of maximum potential energy is unstable.

EXAMPLES

1. A circular disc of mass M lies on a smooth horizontal table; if a particle of mass m resting on the disc is attached to the centre by a spring which exerts a force μx when extended a length x, prove that the period of oscillations when the spring is extended and then set free is

$$2\pi \{Mm/(M + m) \mu\}^{\frac{1}{2}}.$$ [S. 1914]

2. A weightless rod of length 3 feet, with equal heavy rings at the ends, one of which can slide on a smooth horizontal wire, is describing small oscillations in the vertical plane containing the wire. Shew that the period of oscillation is about 1·36 seconds.   [S. 1916]

3. A running watch is placed on a smooth horizontal surface so that it may be regarded as free to rotate about its own centre of gravity. It is composed essentially of a balance-wheel of moment of inertia i, and a body

of moment of inertia I connected by a hair spring. The inertia of the hair spring and other parts of the mechanism may be ignored. Determine the angular motion, and shew that in general, in addition to its oscillations, the body of the watch rotates steadily with a small uniform spin.

If $I = 110$ gm. (cm.)^2, $i = 0.05$ gm. (cm.)^2, shew that the watch gains nearly 20 seconds a day in this position, if its rate is correct when the body is rigidly held.   [M. T. 1920]

4. A uniform bar AB of mass M is suspended from a point O by two equal elastic strings OA, OB. In the position of equilibrium the length of each string is l, and its inclination to the vertical is α. The increase in the tension of either string when its length is increased by x is Ex, where E is constant. The bar vibrates vertically, remaining horizontal. Shew that the period of a small oscillation is $2\pi/p$, where

$$p^2 = \frac{2E \cos^2 \alpha}{M} + \frac{g \sin \alpha \tan \alpha}{l}.$$ [M. T. 1924]

5. A light rod AB, carrying a heavy particle at B, is connected by a smooth hinge at A to a second light rod AO. OA, AB are initially in the same straight line, rotating with uniform angular velocity about a fixed centre O.

AB is now displaced through a small angle, relatively to OA, in the plane of rotation. Shew that, if the angular velocity of OA be maintained constant, AB will oscillate relatively to OA with a period $T\sqrt{l/a}$, where T is the time of revolution of OA about O, and a and l are the lengths of OA and AB respectively.   [M. T. 1927]

6. One end of a uniform rod rests on a smooth horizontal plane and the other end is attached by a string of length l to a point whose height above the table is greater than l. Shew that, if small oscillations take place in a vertical plane, the length of the equivalent simple pendulum is $2l$.

7. A uniform plank, length $2a$ and thickness $2h$, rests in equilibrium on the top of a fixed rough cylinder of radius r whose axis is horizontal. Prove that, if r is greater than h, the equilibrium is stable; and that, if the plank is slightly disturbed, the period of an oscillation is that of a simple pendulum of length

$$(a^2 + 4h^2)/3 \, (r - h).$$ [Coll. Exam. 1913]

8. Three equal uniform rods AB, BC, CD of length a are smoothly jointed at B and C and suspended from points A, D in the same horizontal line at a distance a apart. Shew that, when the rods move in a vertical plane, the length of the equivalent simple pendulum is $5a/6$.

9. Two masses m and m', which are free to move on a given circular wire of radius a, are connected by a light rod subtending an angle 2β at the centre of the wire; find the position of stable equilibrium when the

EXAMPLES 259

plane of the wire is vertical; and shew that, if the masses are slightly
disturbed, the length of the equivalent simple pendulum is

$$a\,(m+m')/\surd(m^2+m'^2+2mm'\cos 2\beta).$$

10. A solid uniform sphere of radius b makes small oscillations at the
bottom of a fixed hollow sphere of internal radius a, the surfaces being
sufficiently rough to prevent sliding and the motion being in a vertical
plane. Shew that the length of the equivalent simple pendulum is $\tfrac{7}{5}(a-b)$.

11. A circular cylinder of radius a surrounds another cylinder of smaller
radius b. The former rolls on the latter, which is fixed. Shew that the
plane through the axes moves like a simple pendulum of length

$$(a-b)\,(1+k^2/a^2)\,;$$

k being the radius of gyration of the moving cylinder about its axis.

[M. T. 1910]

12. A cylindrical shell of mass M and radius b is free to turn about its
axis, which is horizontal, and another rough cylindrical shell of mass m
and radius a is placed inside it; prove that the plane through the axes
will oscillate like a simple pendulum of length

$$(b-a)\,(2M+m)/(M+m).$$

13. A body of uniform density, having the form of a sector of a circle
of radius $2a$, performs rolling oscillations of small amplitude with its
circular edge in contact with a rough horizontal plane. Its centre of
gravity is half-way between this edge and its centre. Prove that the
length of the simple equivalent pendulum is $2a$.   [M. T. 1928]

14. A particle can move in a smooth circular tube which revolves about
a fixed vertical tangent with uniform velocity ω. Find the position of
relative equilibrium of the particle, and shew that the time of a small
oscillation about that position is

$$\frac{2\pi}{\omega}\cdot\left\{\frac{\sin\alpha}{1+\sin^3\alpha}\right\}^{\frac12},$$

where α is the angle of inclination to the vertical of the radius to the
particle when in relative equilibrium.   [S. 1917]

15. A conical pendulum executes small oscillations about a state of
steady motion in which the string of length l is inclined to the horizon at
an angle α: shew that the period of its oscillation is

$$2\pi\sqrt{\left\{\frac{l}{g}\frac{\sin\alpha}{1+3\sin^2\alpha}\right\}}.\qquad\text{[Coll. Exam. 1907]}$$

ANSWERS

3. The period of the balance-wheel is $2\surd(I/I+i)$ secs.

14. $a\omega^2\,(1+\sin\alpha)=g\tan\alpha$, where a is the radius.

CAMBRIDGE: PRINTED BY
W. LEWIS, M.A.
AT THE UNIVERSITY PRESS

Printed in the United States
By Bookmasters